D0794200

More Praise for

Profit from Experience

The National Semiconductor Story of Transformation Management

"A book that I didn't have time NOT to read. It is chock-full of very useful information, much of which I plan to adapt within my agency."
—**Julie Meier Wright**, Secretary of the California Trade and Commerce Agency

" When writers balance hard information with human qualities, it makes for facinating reading. Amelio and Simon have successfully infused humanity into a powerful and practical business book for any manager who needs to strengthen a business or a functional group."
—**Kenneth Gilmore**, former Editor-in-Chief, *Reader's Digest*

" Through Gil Amelio's leadership, a struggling National Semiconductor has been transformed into an industry giant. Now with the publication of *Profit from Experience*, we are fortunate to get first-hand insights."
—**William S. Cohen**, Senator from Maine

"[Gil Amelio] discusses management ideas that have worked in the relentlessly fast-moving chip business."
—**Evan Ramstad**, Associated Press

"There's no boasting about accomplishments here. This book is written in a style that says 'this worked for me and it might work for you. Give it a try.' Our advice is the same: Give this book a try."
—Soundview Executive Book Summaries

Profit from Experience

The National Semiconductor Story of Transformation Management

Gil Amelio,
CEO, National Semiconductor

William L. Simon

VAN NOSTRAND REINHOLD
I(T)P™ A Division of International Thomson Publishing Inc.

New York • Albany • Bonn • Boston • Detroit • London • Madrid • Melbourne
Mexico City • Paris • San Francisco • Singapore • Tokyo • Toronto

I(T)P ™ Van Nostrand Reinhold is a division of International Thomson Publishing Inc.
The ITP logo is a trademark under license.

Printed in the United States of America.

For more information, contact:

Van Nostrand Reinhold
115 Fifth Avenue
New York, NY 10003

Chapman & Hall GmbH
Pappelallee 3
69469 Weinheim
Germany

Chapman & Hall
2-6 Boundary Row
London
SE1 8HN
United Kingdom

International Thomson Publishing Asia
221 Henderson Road #05-10
Henderson Building
Singapore 0315

Thomas Nelson Australia
102 Dodds Street
South Melbourne, 3205
Victoria, Australia

International Thomson Publishing Japan
Hirakawacho Kyowa Building, 3F
2-2-1 Hirakawacho
Chiyoda-ku, 102 Tokyo
Japan

Nelson Canada
1120 Birchmount Road
Scarborough, Ontario
Canada M1K 5G4

International Thomson Editores
Campos Eliseos 385, Piso 7
Col. Polanco
11560 Mexico D.F. Mexico

2 3 4 5 6 7 8 9 10 QEBFF 01 00 99 98 97 96 95

Library of Congress Cataloging-in-Publication Data

Amelio, Gil.
Profit from experience : the National Semiconductor story of transformation management / by Gil Amelio and William L. Simon.
p. cm.
Includes index.
ISBN 0-442-02055-4
1. National Semiconductor Corporation—Management. 2. Amelio, Gil. 3. Semiconductor industry—United States—History. I. Simon, William L., 1930- . II. Title.
HD9696.S44N383 1996 95-35688
CIP

Project Management: Raymond Campbell • Art Direction: Jo-Ann Campbell • Production: mle design • 562 Milford Point Rd. Milford, CT 06460 • 203-878-3793

ACKNOWLEDGMENTS

From Bill Simon—

An old adage says that if you have a job to be done, you should give it to a busy person. A man whose schedule often seems to be blocked in five-minute segments, Gil Amelio still somehow found the time to provide the input and guidance, and to do the manuscript reviews, that made this book possible. He has earned my admiration and appreciation, and my thanks.

It's traditional when receiving an award (think of the Oscars), or writing a book acknowledgment, to acknowledge the spouse and other family members. In the case of my wife, Arynne, the recognition is deserved and given from my heart. She has been a courageous and demanding editor; she does not give in easily and I appreciate that stamina. Not only my intellectual conscience, she also makes life and work a continual joy, and each day a surprise. Thanks also to my daughter Victoria and my step-son Sheldon Bermont for listening to details about the progress and complaints about delays; the loneliness of writing is relieved by patient family members and many good friends.

Mary Ann Phillips, whose title at National is director of corporate communications, in fact is able to come up with any answer or fulfill any request. How she manages to meet all challenges is a riddle to which I will not likely ever know the answer.

In Gil's office, the remarkable Bonnie Murphy, she of the sunny disposition, has been enormously helpful despite the unending phone calls and requests. Her assistant, Stephanie Shero, seems to have inherited the same qualities. Mark Levine also aided in tracking down elusive information.

Steve Sharp, who created the art work, suffered through countless rounds of changes without complaint.

A number of people have lent a hand in reading parts of the manuscript and providing feedback. Leading this list is Rebecca Howard. David Simon, my twin brother, and Professor Tom Steiner, were both helpful and willing. Mike Townsend, a consultant to National and a friend to me, was generous and helpful in providing information and reviewing the manuscript, and merits a special thanks.

Others who gave of their own time to provide input were Walt and Linda Brown, John Osborne on financial matters, and Lieutenant General Don Johnson.

I'm especially grateful to one of the country's leading editors, former Readers' Digest Editor in Chief Kenneth Gilmore, whose insights led me to rethink the story-telling style.

Finally, in my own office, thanks to Jessica Dudgeon for the able help in organizing the research materials, and to Lisa Bennett, for interrupting infrequently.

From Gil Amelio —

My thanks go foremost to Bill Simon, who attended one of my summer lectures at the Stanford Business School and suggested that the material would form the cornerstone for a book. Since I had already been thinking along those lines, the suggestion was very timely. His efforts in getting my management ideas and methods down on paper are much appreciated; this volume would not exist without him.

I am indebted to all members of the Board of National Semiconductor for the confidence they displayed in offering me the position of chief executive officer. Three in particular deserve mention. J. Tracy O'Rourke, chairman and CEO of Varian Associates, Inc., has been a mentor *extraordinaire,* as well as a friend, for the last decade. And Modesto ("Mitch") Maidique, Ph.D., president of Florida International University, has consistently coached, challenged, and encouraged the academic rigor of my thinking.

I'm particularly indebted to Charlie Sporck, personally and professionally. As CEO of National for some quarter of a century, Charlie took the company from nearly zero to $1.7 billion in sales, an achievement matched by few others in American business history. In the process, he created an egalitarian company, where to an amazing degree people are free of playing the destructive games of office politics.

Emory Business School Professor Bob Miles, and National's Richard Feller, have each been valued contributors in the development of the business methods described here, and both continue to serve in that role.

My thanks also to Mike Townsend, whose memory of "how it was" extends back to my pre-National Semiconductor days. Mike's ideas and efforts have helped make transformations happen at each of my last three efforts, and his comments have been valuable in this writing project. And while Mike and Bob Miles began working with me in my early days at Rockwell, the new ideas I was able to pursue with their help at National would not have been possible without the additional inspired leadership of Rockwell International's Kent Black and Don Beall.

My executive assistant, the unflappable Bonnie Murphy, bore up under the added burdens of the book project (and Bill Simon's sometimes daily phone calls) without losing her good-natured smile. Alan Markow, National's new vice-president of corporate communications, also assisted with a review of the manuscript.

And a special thanks to my wife Charlene for her extraordinary patience, support, indulgence, and penetrating comments.

CREDITS

Author photo by Doug Menuez/Reportage, Sausalito, California.

Art work in this book is by Steve Sharp, of Sharp Ad and Design, San Diego. In some cases his designs were based on art that originated at National Semiconductor.

CONTENTS

PREFACE

> *"Few organizations are so weak they can't be saved, few so healthy they can't be improved."*

"THIS IS THE STORY..."

In 1991, National Semiconductor was teetering on the brink of bankruptcy, with cash on hand equal to only three days of receivables.

A Fortune 500 company recognized as one of the world's leading chipmakers, National had lately been in severe decline, reeling from ailments it could not diagnose. The company closed its books for fiscal year 1991, showing income a record $161 million in the red—the worst loss in its thirty-year history. The aggregate loss since the beginning of the downslide in the mid-eighties was soaring toward half a billion dollars.

Three years later, National would report the highest earnings in its history—$264 million on $2.3 billion in sales.

By 1995, the Board would acknowledge Amelio's success naming him not just CEO, but Chairman of the Board as well.

This is the story of the transformation at National Semiconductor, and Gil Amelio, the man who led it.

It's also a marching manual on how to revitalize a failing company or strengthen a healthy one. The principles in this book are a guide for managers and for those who are inspired to lead. Gil Amelio's techniques can realign attitudes and behaviors in the ways necessary to transform an organization. At National Semiconductor, by encouraging his people to take responsibility with him for the future of the company, he has succeeded in more than just typical business profit-improvement: he has transformed individuals into a new style of manager and leader.

In Amelio's words, "Our nation and our industries need people who can understand problems and find new solutions. These are the people who will become the new leaders."

REPLACING A LEGEND

At National Semiconductor, Gil Amelio was taking over from a man who had become legendary. A pioneer of the semiconductor industry, Charlie Sporck was one of the most revered men in Silicon Valley. He was immensely popular within National—the door to his office was always open to anyone in the company, from vice president to production-line worker. (In fact, there was no door to close: Sporck worked in a typical doorless office cubicle like any lower-level supervisor.) He was comfortable as a walk-around manager—roaming the halls and plants, asking questions and making small talk while checking up on everything. They called him Charlie, looked forward to his occasional pat on the back, and took pride in knowing him.

In the fall of 1990, over dinner with their wives, Sporck mentioned to Amelio[1] that he had decided to step down as National's CEO and had told the Board to find someone else. He looked at Gil and asked, "Would you be interested?" Gil said, "Yes," and within days the discussions with the Board had begun.

But how could Amelio replace this legendary figure, this immensely popular leader—a man whose management style was to involve himself in every decision and run the company on the instinct gained from many years of hands-on experience? Their styles were so different that the entire culture of the company would need to change.

---

[1]Charlie Sporck had been concerned over the inroads made by the Japanese, supported by their government-sponsored industry cooperation, and had coaxed and cajoled the heads of U.S. semiconductor companies to join in a similar venture. The result was the Semiconductor Industry Association (SIA), and its research arm, Sematech, a consortium financed by the U.S. semiconductor companies and sharing its technology with the sponsoring companies. Sporck has been referred to as "the father of Sematech." Gil Amelio had been a strong early supporter of Sematech, and had served on the Board of Directors of the SIA. He and Sporck had first met when Gil, at Fairchild, was responsible for a collaborative project with National. But the two men came to know and admire each other through the SIA and their shared concerns for the industry.

TAKE-OFF

On a Thursday afternoon at the end of January, 1991, Gil cleared out the desk of his office in Dallas, where he had been a group president with Rockwell International. Leaving with some misgivings but with the blessings of Rockwell management, Gil and his wife Charlene drove straight to the airport and climbed into the cockpit of his seven-place Piper Cheyenne turboprop. Taking off at dusk, they headed toward the setting sun—toward a future as uncertain in some ways as ever faced by a family of pioneers in a covered wagon.

At 7:30 the next morning, at a hotel in Santa Clara, California, Charlie Sporck was beginning a two-day business review with sixty of his top managers. A low murmur of anticipation electrified the meeting room when Gil walked in. Here was the person hand-picked by Charlie and the Board to lead the company. His reputation as a remarkably successful transformation manager was well known, but what sort of person was he? What could he do for National that Charlie, who had run the company for so long, who had made it into a global, multi-billion-dollar business, had not already tried? And how many of those in the audience might soon be out of a job?

Gil recognized that everyone was waiting for him to ask perceptive questions, or toss out insightful, remarkably cogent observations. Yet the presentations were peppered with indecipherable references—products cited by code name or part number, departments by acronyms, and mysterious, unidentifiable items by cryptic sets of initials. He would later joke that "My greatest challenge in taking over at National was to figure out what all the abbreviations meant."

After two days of reports and discussions, Amelio left with mixed messages and few conclusions. The executives were clearly "a group of very bright, committed people" who were enthusiastic, and making a real and conscientious effort. But the situation seemed even more dire than he had anticipated.

One example was the approach to accounting—which is not, despite common assumptions, the same at every company. When he was later able to review the financials, Amelio's suspicions were confirmed: incredibly, the accounting system shed little light on which products were making money and which were losing. Individual product lines were not being routinely monitored in terms of gross profit; in only one of the thirty lines was the manager routinely calculating and reporting the gross product figure. If an item had large sales, everyone smiled, and no one asked whether it was really helping the company's bottom line.

The National executives, Gil said later, "did not adequately understand the business profit equation or what it means to do a strategic plan, were confused about the current meaning of (markets,) or by delivering value to customers." Even worse, they were a management group "without a clear guiding philosophy of management."

Amelio realized that he had underestimated the problems he faced, and would need to completely rethink how to handle the challenges ahead. He had achieved stunning successes in the past. But now he wondered, "What the hell have I gotten myself into?"

THE FIRST FOUR MONTHS

The Board had agreed to give Amelio a four month breaking-in period, during which he would have the title of "co-CEO." While Sporck continued running the company, Amelio would be free to walk, talk, travel, visit, pore through files, and ask probing questions. By the end of the process, he had formulated a picture of what was right with National Semiconductor, and what was wrong with it.

The greatest strength by far, he thought—and looking back he credits this to Charlie Sporck—was a solid work force. National had a lot of gifted, productive people who had stayed with the company through the dark years.

On the down-side, Gil believed the company lacked a clear vision of the future. They were trying to do too many things—striking out in a variety of directions. The attitude was seat-of-the-pants, left over from the company's go-go early days—"If we might be able to make money on this, let's go do it."

Some companies survive, maybe even thrive for a while, on such an approach. But Gil saw that the business practices at National hadn't kept up with growth nor with changes in the industry, and were fundamentally out of step with the needs.

Facing problems of this magnitude, some new CEOs might have given up on the spot and sent out their resumes.

It was May, 1991. For Silicon Valley, another glorious California summer was just ahead. For National Semiconductor, the weather forecast was "cloudy, conditions uncertain."

IS THIS BOOK JUST FOR CEOs?

Gil Amelio's management successes at National and earlier are based on straight-forward tools and principles—not theories, but specific, down-to-earth approaches that any chief executive can apply on a company-wide level... or that managers and supervisors can apply to revitalize their own organizations and improve their own management style.

In these pages you'll find those techniques spelled out, with an emphasis on the ways of getting employees committed to transformation, and the financial measures that really tell you if your organization is succeeding

Amelio says, "Few organizations are so weak they can't be saved, few are so healthy they can't be improved."

PART ONE—PEOPLE LEADERSHIP

SECTION 1

Charting the Course

1

Creating the Vision

*"To achieve success, you must first **define** success."*

Approaching the door to a conference room, Gil Amelio is surprised and puzzled by the unbusiness-like sounds of nervous laughter from within. He's about to spend the morning listening to presentations from a group of National Semiconductor managers, and sharing his transformation message with them. He's done this before, on a regular basis, but there's something ominous about what he's hearing.

After polite introductions and opening remarks, Gil discovers the reason for the electric tension in the room. From offstage, several managers enter to deliver the group's formal presentation. But instead of appropriate business clothes, they're dressed in outlandish costumes.

The uncertainty clouding the room is, after all, justified—people get fired from companies every day for misjudging the disposition of the boss. Gil has showed himself to be serious but not stern, demanding but not unfair, willing to grin and use humor—but only in appropriate settings. These managers don't yet know him; how will he respond to this unconventional approach at an important business session?

Gil's quizzical expression suggests he's withholding a reaction, willing to give them a chance to demonstrate that their motives aren't frivolous. That's a relief—maybe it will be all right after all.

The reason for the costumes quickly becomes clear. As part of their week of sessions, the managers have put in long hours working on their presentation. They have some changes to propose for the company's new Vision statement, and a series of important recommendations about the transformation. But Gil has already attended a dozen or more of these sessions; will he still be open to further input? The costumes are an attempt to bridge that hurdle by setting an unanticipated mood.

One of the costumed men begins to speak. He is marketing manager Brad Paulsen, from National's Mil/Aero group. Sporting a scraggly gray beard, Paulsen holds a cut-out in the shape of the ancient tablets. He is Moses.

The presentation is called "Moses and the Profit$."

Some of the managers have had cold feet about the gimmick; now they're relieved to see Gil beginning to respond positively—open, smiling, listening attentively, fully focused, even nodding in acceptance.

Gil has recognized that this is going to be a more valuable session than most. It demonstrates that the effort to set National Semiconductor on a new course is not only taking hold with the employees, but even generating enthusiasm, creativity, and spirit. It's spawning some productive ideas, some positive feelings, and even some laughter—signs that the task of saving the company might not be hopeless, after all.

PRIORITIES

I was convinced almost from the day I walked in the door to become National Semiconductor's new CEO that the top management team showed both strength and quality. But how did an organization get into trouble with so many valuable contributors?

The answer lies in lack of focus. Capable people are essential, but not sufficient; to be viable and successful, a company must have a clear course, clear messages, and clear processes, all clearly communicated.

White light shines in every direction and into every corner—great for illumination, showing only the surfaces. With laser light, all the energy is focused to create a tremendous intensity—enough to cut through to the interior and reveal the core.

National needed leadership and direction. But most of all, National needed laser-like focus.

Anyone familiar with the recent lore of American business knows the story of Jack Welch's cultural revolution at General Electric—a useful benchmark for understanding the transformation at National. But where Welch took on a robust, money-making company, I had assumed leadership of an organization that was, as I've sometimes irreverently described it, "less than a week away from debtor's prison."

On the other hand, Welch faced the tremendous challenge of convincing people that GE had a problem. At National, everyone was painfully aware that their company was no longer the industry leader it had once been. I sensed they were ready to respond to leadership, willing to find out what unexpected solutions I would be able to pull out of my briefcase. Yet they were justifiably anxious about whether any solutions could work fast enough to save their jobs and their pride.

DON'T CALL IT "TURN-AROUND"

Magazine writers have described me as a "turn-around manager." It's a label I thoroughly dislike, a "hot button" that launches me into a super-charged explanation of the difference between "turn-around" and "transformation."

A turn-around manager is an executive who, when he's brought in to save a troubled company, follows a ruthless formula. In the extreme case this involves slashing all budgets not directly involved with generating revenues, decimating the research expenditures, selling off a few operations to raise cash, and laying off half or more of the work force.

These quick-fix solutions don't address any of the business disciplines or management problems, but only hack away at every corner that might produce a fast improvement in the profit picture. This creates the appearance of health, and the manager is hailed by shareholders and the analyst community for his remarkable achievement.

He soon departs, significantly richer, headed for his next position as a short-term hero somewhere else. What he leaves behind is a company no longer able to function productively, a company without the research to create new products, without the staff to support a marketing effort or provide customer service—in short, without any means to sustain cash flow and to grow. The extreme turn-around manager is a pirate who sails away leaving a floundering crew on a plundered ship heading toward disastrous storms.

A *transformation* manager looks for long-term solutions; he or she intends for the company to become strong, healthy, and successful, and to exist and thrive for many years.

Shun the easy turn-around. Follow the road less traveled: be a transformation manager.

THE VISION: A DEFINITION OF SUCCESS

The current business dictum says that every company must have a Vision statement. You're supposed to have one, but does it really make a difference?

On joining a company or moving up to any higher management position, it's natural to assume that your people have a clear image of where they're heading. Don't be deceived—ask a few of your workers what the organization is specifically aiming for. Even when there are copies of a Vision statement hanging in every corridor, the odds are the employees don't know and cannot tell you, or think they know but cannot agree.

Try the same thing with your customers. If you want customers to be buying on something other than price alone, they should be able to say what differentiates your company from the competition. The odds are that the way they describe you doesn't sound anything like the company you want to be.

To be successful, you must first *define* success; to reach a goal, you must define that goal. When others are involved, you can reach a goal only by first reaching an understanding and agreement on the definition.

When I arrived at National, I set about doing the things I had done in previous positions. One of my early priorities after coming to grips with the financial emergency (more on this in the Financial section) was to set in place a corporate Vision, which would serve to provide both the definition of success at National, and a description of our goal that would get everyone started working in the same direction.

When sailing a small boat across a bay or lake you know well, you and your crew members don't need to rely very heavily on maritime charts and weather forecasts—you can spot the navigational markers and judge the clouds just by paying attention as you go. But when you intend to set off from San Francisco bound for Hong Kong, you'll chart a course and thoroughly brief the crew before untying the dock lines.

In business, the definition of success needs to be clearly charted; it should not be left for the employees to make assumptions. Every dysfunctional organization I've ever studied has shown this lack: the corporate "crew" could not agree on nor clearly articulate the company's vision, and individuals couldn't even correctly describe the intentions of their immediate boss.

A business, unlike a sailboat, leaves no choice of when to cast off the dock lines—since the business is in mid-voyage, you need to chart the course with one hand while steering with the other, and remain ready to tack when the wind changes.

Any organization—whether 50 people or 50,000—becomes more powerful when everyone understands where they're heading.

"Vision" has become an over-used corporate buzz-word of the '90s. Many companies have a Vision statement posted the way Stalin's picture used to be posted in the Soviet Union—and, I think, with as little benefit. Before we get bored with the "vision" word and toss it aside, let's recharge it to our benefit. Realize first that just publishing a Vision statement and speaking in visionary terms at your communications meetings doesn't automatically enroll your people to accept it, believe it, and make it work.

So how do leaders add new muscle to the faded concept of Vision; how do you encourage your people to respond to a "new, improved Vision?"

VISION, MISSION, AND PURPOSE

Because people sometimes ask for clarification of terms, here are my own distinctions: Vision is a statement of what the company wants to become; Mission is a more precise description of what the company does—which businesses it should be in and which not; and Purpose is associated with specific objectives—to double sales annually, or gain five points of market share.

WHAT GOES INTO A VISION STATEMENT

From the design concept that a naval architect creates for a new ship, detailed plans will be drawn. Eventually every craftsman in the yard will follow precise specifications that flow out of the original concept. In the corporate world, the design concept is the Vision.

Your Vision statement, expressing *what you are trying to become*, must be expressed in terms of something that can be grasped and internalized, something people can identify with.

But it must go beyond this. It becomes strategic by incorporating an expression of *values*—the values you share throughout your company, and the value you deliver to the customer.

Apple Computer's Vision statement, "Changing the World One Person at a Time," offers a powerful vision; because Apple was not then undergoing transformation, the statement did not need the additional strategic elements. The *New York Times* Vision, "All the News That's Fit to Print," is another strong statement, but again without any values.

Notice how National Semiconductor's Vision statement becomes a *strategic* Vision by including values:

Vision of National Semiconductor

> A distinctive, exciting company, valued by its shareholders, employees, and customers, respected by its competitors, and acknowledged by the rest of the business world as a leader, a major asset in the communities in which it serves and operates.

STRENGTHS AND WEAKNESSES

At National, my first four months were the springboard. This period as co-CEO had given me the opportunity to take the pulse of the company and begin to gain a sense of the direction in which we needed to be moving.

As I had recognized from the first, the financials were not organized in a way that revealed the profit profile. Because manufacturing had been an area of Charlie Sporck's expertise, I had anticipated this would be an area of major strength but it was soon clear the reality was very different; fixing the problems in manufacturing proved to be a major hurdle.

On the other hand, engineering—which I had expected to find rusty and second-rate—proved to be doing a lot of top-quality work.

On the "soft" side of the equation, each plant I visited held a large communications meeting so the workers could get a glimpse of the new CEO—often a thousand people or more, all of them afraid for their jobs. Few things a CEO does are more challenging than trying to establish rapport with a large group in thirty minutes. If there is a secret to this, it probably lies in believing, as I do, that *people* are the path to success for any company—which is not as obvious as it sounds, because the tendency is to proclaim this belief while actually putting your reliance just on the technology or the numbers.

I recall visiting one of the National plants in Malaysia. There wasn't a sound in the room but my voice. The people sat there with such passive expressions that I suspected, despite reassurances, that they didn't speak enough English to understand me.

Yet afterward, dozens of people came up to share a few words of encouragement or tell me they were excited about the changes coming and were 100 percent on the team. These thoughtful, caring, loyal people added to my short list of reasons for confidence and hope.

The management team, on the other hand, left a sense of desperate concern. More competent as engineers than as managers, would their dedication and hard work be enough for the crew that was to sail the National ship into the future?

THE EXECUTIVE OFF-SITES

In June, 1991, as one of my first principal actions after Charlie had left and I had assumed full authority, I took the top forty executives of the company to a facility a short way down the coast in Santa Cruz, California, for an intensive half-week—the first of what would become regular, quarterly Executive Off-sites.

I intended this to be an analysis of the company and, at the same time—but not obviously, I hoped—a tutorial on the management skills that would be needed. My fear was that this management-development effort might not be effective quickly enough to keep the company alive.

Throughout the early part of my remarks, the group was impassive and unresponsive. Clearly, I wasn't connecting.

It turned out they were wondering how they should react—not wanting to speak up until certain of what behavior was acceptable. They were being confronted with the first step in a dramatic change in culture—from the regime in which Charlie Sporck reserved to himself a voice in all the key management decisions, to a new era in which I would expect, not just participation, but responsibility and decision-making pushed further and further down the line.

My initial alarm subsided as the group progressed rapidly beyond the boss/employee mode, to become a team of colleagues engaged in a common quest. By the end of the session, I knew National could survive if together we could move rapidly enough.

Many of the topics I've given the executive management group to wrestle with at the Executive Off-sites are topics that will be addressed in these pages. A high priority was to elicit their participation in designing a corporate Vision statement—to identify National values and to define what we wanted our company to become.

CREATING NATIONAL'S VISION: FIRST PASS

Canadian poet laureate Bliss Carman once wrote, "Set me a task in which I can put something of my very self, and it is a task no longer; it is a joy, it is art."

Involving the management team in this way would give them an opportunity to become a crucial part of the transformation process. And it would be a key to gaining their support.

I launched the Vision discussion at an early Executive Off-site by talking about the role of a Vision statement and what it would need to accomplish. In preparation, I had brought along my own draft of an initial version as a jumping-off point. The group then broke into teams to discuss the draft, poke holes in it, and shape something they would find more suitable. The text went through numerous versions over the next couple of days as they enthusiastically hammered out a wording they could subscribe to without hesitation.

That my efforts with the group were beginning to pay off became clear when one of the senior executives afterward commented, "For most people, Charlie would have been a tough act to follow."

ENGAGING THE RIGHT BRAIN

A Vision statement for an organization in transformation should appeal mostly to the right brain. It should strive to be uplifting rather than excessively practical.

The much-praised Jack Welch statement about being a market leader in each product line ("Be number one or number two") is, in my view, too mundane.[1] With the benefit of hindsight, Welch might agree that this makes you want to break out the market studies and see, yawn, how close you're coming. "World leader," on the other hand, sets a goal people can get excited about.

Business professor Richard Ellsworth, who has been watching GE for twenty years, shares this view. He perceives "a certain hollowness of purpose" and told *Time* magazine, "...what (Welch) hasn't articulated is the reason why they are competing, a more meaningful set of values. He has not given GE a morally uplifting tone."[2]

This appeal to the right brain is one of the most important underlying values of the approach I advocate.

If your Vision statement is bland, you will not get people to make a commitment. A Vision statement should stir people's souls. It should set higher sights. It should elevate.

CREATING YOUR OWN VISION STATEMENT

It's easy to think that anything so formal as a Vision must only be for giant corporations. On the contrary, I would put it as one of the fundamentals for anyone who wants a more effective organization. I'll draw a line in the sand: anyone who is reading this book and works for an organization that lacks a competent Vision statement should create one for his or her part of the organization, and should be promoting the idea of a Vision statement for the company as a whole.

The Vision statement you're going to create will be designed along the same lines ours was—to present a description of the future you aspire to achieve.

I can imagine someone reading this and feeling the stress beginning to build—thinking, "I wouldn't know where to begin."

Picture yourself as a reporter from *Fortune*. It's five years from now, and you're writing a story about your organization's success. Try it—write the lead paragraphs, or outline the three or five key points you'd like to appear in the story. How would you want the headline to read? What would you want the article to say?

Another approach: consider the key phrases in the National Semiconductor Vision statement—

- Distinctive, exciting company
- Valued by its...employees and customers
- Respected by its competitors
- Acknowledged as a leader

If you're having trouble getting started, pick the most appropriate one or two of these, and see how you would mold and adapt them to reflect the goals and values you hold in your own organization.

For a new organization, set your sights on a relatively short-term horizon—three years, if you can; if things are evolving so fast that you cannot see three years ahead, make it two, or even one.

Develop your Vision statement through a contributory process. Get involvement from a wide group, try out drafts on stakeholders, find out which statements are clear and meaningful, rework the whole until it's conveying a valuable message that will provide motivation and guidance to the entire organization.

OTHER PEOPLE'S VISIONS

While many companies have not created a powerfully *short* Vision statement corresponding to National's, here are excerpts I admire from the direction-setting documents of other U.S. firms, which may be helpful in your Vision-creating efforts:

"We are committed to our People-Service-Profit philosophy...We will strive to have a satisfied customer at the end of each transaction."

Federal Express

While putting people first, the company nonetheless acknowledges that profit is a motive.

"... (A) company that our people are proud of and committed to, where all employees have an opportunity to contribute, learn, grow, and advance on merit...."

Levi Strauss & Co.

The Strauss statement devotes specific sections to the issues of diversity, and ethical management practices.

"Create the most powerful, highest quality computational tools for solving the world's most challenging scientific and industrial problems."

Cray Research

Although addressing only the business goal, Cray's statement sets a Mt. Everest target.

"We believe our first reponsibility is to doctors, nurses, patients, mothers, and all others who use our products and services..."

Johnson & Johnson

When several people in Chicago died after taking Tylenol™ headache medicine, Johnson and Johnson didn't wait to find out the cause (which turned out to be not problems at the plant, but tampering with bottles on retail shelves). The company ordered an immediate, total recall of the product, nationwide. This decision was guided by the quoted sentence, the first line of the Johnson & Johnson Mission statement.

A VISION FOR EVERY WORKGROUP

Once your organization has an overall Vision statement, you've established a pattern that can then be followed throughout all levels of management.

Executives and managers, in conjunction with their employees, create a Vision statement for their organization that supports the main Vision of the company.

AN EVOLVING DOCUMENT

Our Vision is, as yours should be, a living concept, continually being re-evaluated and amended. As conditions change within the company, as the marketplace and the technology evolve, the organization must evolve in response, and the Vision must change to reflect the new conditions.

In June, 1995, at the time of writing this, I am preparing for our next Executive Off-site, where one major agenda topic will be how our Vision and values need to be changed to accommodate a company that has returned to health and is now targeting "greatness."

WRAP UP

Is creating a Vision worth the bother? A study[3] by Professor Jerry Porras of the Stanford Business School followed eighteen "visionary" companies that set goals, communicated them to their employees, and had an image beyond the obvious one of making a profit (a prime example: Disney's "making people happy").

The visionary companies outperformed a control group by more than 600 percent.

So, yes, it's worth the bother. But a Vision statement can be words on paper, issued with hoopla but quickly gathering dust. To make sure your Vision avoids that too-common fate, you want to get disciples throughout the company who will begin preaching the religion on their own and spreading the word that will continue to replicate itself all the way to the grass roots. You want everyone to become not just a student but both student and teacher. Doing this is possible, and not so difficult as it might sound; it's the subject of the next chapter.

The WSJ Takes a Tongue-in-Cheek Look

Like almost any other business tool, the Vision can be worthless if misapplied. *Wall Street Journal* staff reporter Gilbert Fuchsberg finds an element of absurdity in the way a lot of companies approach the Vision statement. Every Marriott hotel, he says, has a different one, Avis has 150 of them, and the august Stanford Business School spent more than a year hammering out theirs—only to produce one virtually the same as the statement it was replacing.

Shortly after taking over at IBM, Louis Gerstner announced, "The last thing IBM needs right now is a Vision." The remark created such a stir that Gerstner saw the light, and issued employees a list entitled "IBM Principles"—in other words, a Vision statement.

END NOTES

1. GE uses what they call a "Shared Values Statement." The 1992 revision incorporated these main points (among others):

- Understand accountability and commitment and be decisive. Set and meet aggressive targets with unyielding integrity.
- Have a passion for excellence. Hate Bureaucracy and all the nonsense that comes with it.
- Have the self-confidence to empower others and behave in a boundaryless fashion. Believe in and be committed to Work-Out as a means of empowerment. Be open to ideas from anywhere.
- Have, or have the capacity to develop, global brands and global sensitivity and be comfortable building diverse global teams.
- Stimulate and relish change; not be frightened or paralyzed by it. See changes as an opportunity, not just a threat.
- Have enormous energy and the ability to energize and invigorate others.
  (As listed in Control Your Destiny or Someone Else Will, Tichy & Sherman, Doubleday, 1993.)

2. As quoted by John Greenwald, "Jack in the Box," Time, Oct. 3, 1994, pp. 56-57.

3. As cited by Gilbert Fuchsberg, "Visioning Missions Becomes Its Own Mission," *Wall Street Journal*, January 7, 1994.

2

Involving Everyone in Defining the Organization

"The uniqueness of the National approach lies in… actively engaging all employees in the process of defining what kind of company to have."

In 1974, on one of those great-to-be-alive days, Gil Amelio was at the tiller of a 27' Coronado sailboat that was clipping across San Francisco Bay. Tending the jib sheets was the boat's new owner, Gil's friend Frank Schwarb, a retired Navy commander.

The boat was roaring along making hull speed in 20 knots of wind. This was what every sailor dreams of—filled sails, a trim boat that handles well, bright weather, a stiff, steady breeze. In waters they both knew well, in such perfect conditions, the two men had no need of charts or weather broadcasts.

They sailed up the bay past San Francisco and around the Garden State side of Alcatraz. And then, with a terrifying noise, the boat lurched wretchedly. Water began slowly rising above the decking. They were sinking. And they were out of control—there was no response to the rudder. The wind was pushing them toward the rocks of Alcatraz.

Frank and Gil struggled into life jackets, each mentally running through their options. The boat was not equipped with a radio—a mistake that Frank would soon remedy and that Gil would never make again. But they managed to flag down a passing craft, whose skipper relayed an emergency call to the Coast Guard and stood off to assist.

By now the boat was so near the rocks of Alcatraz that the other skipper was unwilling to risk coming in close. Frank grabbed the bow line and plunged into the water, determined to swim the distance to the other boat. Gil thought this foolhardy, but Frank hadn't asked for any opinions. Now Gil could see that Frank was having trouble—the other boat was farther off than it looked.

But Gil wasn't doing much better. He had to wrap his arms around the mast as the boat began smashing against the rocks.

The water in the boat was almost knee-deep by the time the Coast Guard rescue boat and helicopter arrived. They fished Frank out of the sea, then came in for Gil. One sailor threw a rope; Gil was afraid if he let go of the mast to grab it, he might go overboard.

Talking aloud to build his courage, moving cautiously, he nabbed the rope and cleated it, and the Coast Guard was able to tow him off the rocks and to a shipyard.

Both men had managed to escape unharmed, and the boat's damage was repairable. The accident had happened because of a submerged rock, unmarked by any navigational buoy or warning device,[1] but clearly marked on the maritime charts of San Francisco Bay.

Gil would never again set sail without charts that he regularly consults. But the experience carried a business lesson as well: the need to involve the crew—the people you lead—in setting the course of your voyage together.

When Amelio was a group president at Rockwell International, he served on the corporate management team. Rockwell decided they needed a corporate Vision, and gave the job of creating one to the vice-president of HR. As the work progressed, drafts were sent around to the top executives for comment. Perhaps because, as so often happens, the individual comments contradicted each other, they were ignored.

The statement subsequently adopted was perceived to be on target and addressed some of the important issues. It was printed up and passed around, but Gil remembers that it didn't seem to offer any inspiration and didn't gather much interest.

Looking back, Gil thinks the reason is clear—no one, not even the top management team, sensed that any ideas, beliefs, or values of theirs were reflected; none sensed they had contributed. With no involvement, they did not feel they had any stake. Gil admits that he did not conscientiously plan or work to reach the goals set by the statement.

What's more, many Rockwell employees shared the view that management too often said things one way but did them another. The situation was compounded when Rockwell printed up thousands of copies of the statement to be handed out and tacked up on the walls before getting any "buy-in."[2]

From experiences like these, Gil eventually concluded that the only reliable way to gain support throughout an organization is to involve as many employees as possible in the process of ***defining*** *the organization.*

Men like Carnegie, Rockefeller, Ford, Steve Jobs, and National Semiconductor's Charlie Sporck were able to build great organizations because they had a clear grasp of where they wanted to go and were somehow able to ***imbue*** *others with their vision—even when they couldn't articulate it. Most managers have ideas and direction; it's the communicating and the "imbuing" that's difficult.*

GATHERING MANY CO-AUTHORS

Despite the continual struggles over products, promotions, and profits, every business is fundamentally a social structure, very much like a small city. A social culture grows out of this; people respond in particular ways, depending on the shared values of the organization. The better they understand those values, the more effectively the community works.

This turns out to be an essential element of creating a corporate direction that people willingly align themselves with, especially when the direction of the company is being changed—when a transformation is taking place.

You must not carve some words onto tablets and pass them around. The goal is to make everyone a coauthor of your corporate dream so that you will get their buy-in. This is an on-going essential ingredient for smooth transformation, and it requires that you involve your employees from the start. It means they will help define what kind of company or organization you will have.

I've described how my top management team worked together to create the initial version of our new Vision statement. The primary purpose of that process was to gain support. The ideas of the management team, based on their knowledge of the company, were valuable—but their involvement was essential. Involvement is the key to understanding and buy-in.

While the people you have daily contact with are easy to reach with your messages, getting buy-in from people beyond your direct reports is a different story altogether. At National, we created a change-enhancing program that we call "Leading Change" (see Chapter 7)—first for the top-level managers throughout the world, and subsequently moving down through the organization.

Earlier I mentioned the idea of assembling material for a *Fortune* article five years from now. We presented this challenge to every one of the thousands of managers and supervisors going through Leading Change. They had the opportunity to examine the corporate Vision, dissect it, and think about what they would change or add. In the *Fortune* exercise, people were fascinated to consider what kind of company they wanted. They treated it like a game; some produced a list of points, others wrote a whole magazine story.

This wasn't just a drill with no end-product. We used many of these ideas in amending and enhancing the Vision. The employees were actually helping to define the company they wanted to work for. And they still are. Though the changes these days are much less substantive, we continue to listen and to make refinements.

Most of the input from the lower level and factory level employees has reflected a concern for the worker's personal life and the balance between work and family. Is that a valid element for a corporation to include in its Vision?

The Gil Amelio answer is, "Absolutely."

We incorporated these concerns in the wording that says we want to be a company that is "valued by its... employees."

Not every worker, not even every executive these days, is ambitious and career-minded. But when you acknowledge the values of your employees, the results can be surprising: we find that the worker who takes a morning off to attend a child's soccer game comes back and puts in more of an effort to make up. As one way of providing the requested help integrating work and family life, we launched an annual "Kids to Work Day," which gives youngsters a context for understanding when parents talk about their jobs and brings in more complimentary mail than any other program we've ever done. In a Community Care program, teams of National volunteers are doing charity work together; one effort, led by administrative assistant Sharon Birks (now a marketing communications specialist) and others, brought forty employees from corporate headquarters together in their spare time over several weeks to refurbish the home of an elderly ill woman, even installing a new roof, as part of a nationwide "Christmas in April" program.

National might have missed such ideas without input from employees. Including "valued by its employees" in the Vision statement announces to everyone that efforts which make the company more valued in the eyes of the workers are desirable and encouraged.

TRANSFORMATION AS A PROCESS

While most people are sure they know exactly what a process is, many, in my experience, have only a nodding acquaintance with what the term actually means in the practical, everyday world of business.

Even the simplest of tasks must have a process—something we try to teach our youngsters at the finger-painting phase: Step one, prepare the tools—brush, paper, paint, apron, newspaper to cover the table. Step two, paint. Step three, clean up.

To be carried out effectively, an idea or strategy needs to be supported by a precisely specified series of actions taking place over a precisely specified period of time. Each incremental action must build on the one before and lead logically to the next. The process plan describes when and how each of the steps will happen.

Transformation has little chance of success when it takes the form of isolated programs. Instead, transformation needs to be a coordinated *process*—often with many of the steps taking place simultaneously, but all coordinated in a logical and well-thought-out series.

Not just for transformation, but with projects of any kind, managers need to encourage their people to revere the power of process. When someone comes to you with a suggestion, insist they have a time-line—a process—with some rough details in it.

Will Rogers is known for saying, "I never met a person I didn't like." We'd all do well to be recognized as people who never met an idea we didn't like. But ideas aren't worth much without a process to turn them into reality.

VISUALIZING THE DREAM

In the process of creating the Leading Change course, the development team called on National consultant Michael Townsend, who, after working through the accumulated notes and talking at length with me, wrote a twenty-page White Paper explaining the company's new Vision. Recognizing that this was still far too long for people to grasp readily, the team brought in a renowned graphics facilitator, David Sibbet, of The Grove Consultants. A sheet of paper stretching several feet was taped along the front wall of the room. Then the team opened up the conversation and fired back and forth the Vision ideas I had originally expressed to them. As these ideas were tossed around, Sibbet translated them into a creative visual rendering. The result was something that is, as far as I know, entirely unique—an artistic conception of our corporate Vision. In the Leading Change workshops, the Vision artwork attracts and holds the participants' attention.

It's not just that the visualization let people grasp the ideas and values; from the very start, the Leading Change participants agreed they found it much easier to make suggestions. Adding an idea to a chart was less daunting than trying to write a paragraph or two on each suggestion. Out of a static process, we had created something akin to an interactive game.

In time, about seven hundred people contributed to the graphic rendering of the Vision. And, as with the Vision statement itself, the artwork is continually being amended and updated. For example, the company itself is symbolically represented in the art as "Spaceship National Semiconductor." In the version of November, 1992, the spaceship is on the ground, being fueled by employees; the fuel is identified as "core competencies." By October of the following year, a new version of the drawing showed the spaceship as having lifted off, soaring past a place identified as "break-even."

When I visit National offices around the world, I see reproductions of the Vision artwork that people have framed or tacked up in their cubicles as a reminder of what it is we're trying to do with this company, what's important to us, where we're trying to take it. Having your own Vision graphic is the "in" wall decoration around National. And many of the people who have put up this artwork are those who made some contribution to it.

In addition, a number of the business units within the company have used the same approach to depict their own, local Vision.

In most cases, each organization has also created a distinctive Vision poster. The poster produced by one unit of the Analog Division features a crown, representing "customer delight." The Human Resources poster shows one globe representing where HR is today in terms of structure and capability and as perceived by the rest of the company; a broad avenue shows people running toward another globe, representing the future. Standing on this second globe are people of many different nations, cultures, and races, symbolizing the diversity of the company's workforce. And they are holding up a victory banner that represents achieving victory by working together.

A SENSE OF THE PAST

The graphic depiction of a company's Vision provides a place for elements that the constraints of brevity prevent from being included in the written version. Our drawing includes elements that, for example, provide a link to the history of the organization. Just as a major university becomes an "institution" after it's been in existence long enough, so a connection to the past gives a better sense of stability within a company.

A new Vision statement always implies the discomforts of radical change. By providing a link to the past, you will help your people feel more comfortable about accepting change.

WRAP UP

There is double value in the effort to involve as many people as possible in the design of the new Vision. In addition to gathering and extending valuable ideas, the process engages employees to think about the company in productively similar ways. Efforts that are synchronized and cumulative provide focus—and focus is essential for a successful transformation.

When transforming your organization, you are aiming at harnessing the energy, intelligence, and pride of your people as quickly as possible toward a common goal. You need to ensure that everyone shares the same understanding of what that goal is.

The uniqueness of the National approach lies in getting so many people involved—not printing up the CEO's Vision and handing it out, but actively engaging employees in the process of defining what kind of company we want to have.

Some say that remaining number one is tougher than becoming number one. Having been in both situations, I don't agree. But transforming a company is a tough, dangerous, exciting adventure—yet eminently doable.

Many leaders are delighted to take all the credit for their organization's new-found success, attributing it to their "natural talent for leadership." In my view, the true measure of success is whether a process can be taught and replicated. The transformation process we used at National is clonable and teachable; that is what makes it valuable.

END NOTES

1. The Coast Guard rescuers told the men that they rescue one or two boats every year that have struck the same rock. Today, 20 years later, Gil goes sailing only rarely, but whenever he is out on San Francisco Bay, he looks for the rock. It is still there, still shown as a hazard on the chart, still unmarked.

2. For a discussion of changing behaviors first and publicizing afterward, see Wilkins, Alan, Developing Corporate Character, Jossey-Bass, 1989.

3

Setting the Initial Vector

"No business ever got fixed by focusing just on the problems. You cannot save your way into success."

"THE CHARTREUSE STRATEGY"

In early 1983, Don Beall, then COO of Rockwell International, and Kent Black, president of Rockwell's Electronics Operations, offered Gil Amelio a challenge—to take over a Rockwell division that was in trouble and see if he could straighten it out. The job would represent a new dimension for Gil—jumping from GM of a Fairchild Camera division, to running the equivalent of a self-standing company with $120 million in annual sales, and manufacturing plants in California, Texas, and Mexico.

Although he knew the division was in trouble and had been through two presidents in two years, Gil accepted the challenge. What no one told him was just how hot the hot water was. He was to learn that the division was losing close to $2 million a month. Much later he would find out that Rockwell had already taken steps to write off the whole operation, with funds neatly stashed in a reserve account specifically earmarked for closing the division. Rockwell management considered Gil the last-ditch attempt; Beall fully expected him back within two months to advise that the unit was unsalvageable.

Meanwhile Bob Miles, then a Harvard Business School professor serving as a Rockwell International consultant, received a phone call from Rockwell. "You'll probably hear from someone named Amelio," the caller said. "Stay away from it. God himself couldn't save that business."

This was to be Gil's first exposure to an operation desperately in need of transformation, but it didn't take a crystal ball to recognize that changes, big changes, would have to begin immediately.

How do you get a quick grasp on a large organization that is unknown to you? One step he took out of desperation, and has used ever since after seeing its value, was a way of sizing up the organization and the people at the same time. He announced a business review, to be attended by all the executive staff. Each of them presented at length, in front of the assembled group, how his part of the business was doing. They broke for lunch, and then went back to it. When the last presentation was finished, Gil stood up and told them, "This division is losing money so fast that every employee is in danger of being out of a job, yet I've been listening to you for hours, and I have yet to hear about a single problem."

It was symptomatic that each executive focused only on his own area, and what was going well; no one spoke about issues involving the working relationships among the different parts of the division, or overall strategic issues that would determine their future, or what was wrong, or what needed to be done to fix the problems.

Some of the transformation techniques Gil now uses were honed at Rockwell in the intense pressure cooker of fix-it-now-or-close-it-down.

From the outset he recognized one aspect that is often overlooked. Companies undergoing transformation talk about change, many achieve it, but few manage to change the mindset of the work force—to get people believing that the changes are going to improve the work atmosphere, the job security, and the attitudes of their co-workers. When Gil Amelio speaks of "setting the initial vector," what he's talking about is a process that gets people's attention.

At Rockwell, he changed the name of the organization from the Electronic Devices Division, to the Semiconductor Products Division (SPD). He took steps to change the culture from the combativeness of separate fiefdoms (a practice that executives at National Semiconductor would later come to call "silos") to the communication and cooperation of a mutually supporting army. While he was changing the business plan, the product mix, and the entire business orientation, Gil was also making sure that people at every level were sensing the new attitudes.

Trying to explain what was going on to a newspaper reporter, Gil said, "I was willing to try anything. It didn't matter if I had to paint the building chartreuse to change the pervasive attitude..."[1]

The phrase was picked up within the company, and Gil's transformation program at SPD came to be known for a while as "the chartreuse strategy."

In three years, revenues had recovered sufficiently to carry the division from a $17 million loss, to a profit of $12 million, which was $5 million over plan. The following year, profits reached a dazzling $34 million—representing a $50 million transformation in three years.

SPD became the world's largest producer of the modem electronics that comprise the heart of a fax machine. At a time when other U.S. electronics firms were complaining that the Japanese market was closed to American products, 70 percent of the fax machines produced by Japanese companies carried SPD boards. Today, more than a decade later, SPD still holds the dominant position globally in this product category.

ACTING BOLDLY

The previous chapters dealt with launching transformation, a process that can be likened to setting a vector. But as you might remember from your physics or math courses, what's special about a vector is that it has not just direction, but also magnitude. An arrow on a weather map can show that the wind is from the northwest; the length shows that it's blowing at twenty knots.

Establishing a vector for a company means more than just setting a direction; it means establishing a new course in a bold way—bold enough to send a message.

At IBM, two years after becoming CEO, Lewis Gerstner had still not made any bold strategic moves or articulated a new visionary course—which the business press took delight in pointing out. He had launched some drastic cost-cutting measures, including another round of layoffs (regrettable but apparently necessary, though not well received at a company where layoffs had long been unknown), but these are plainly not actions to galvinize and motivate the troops. It takes bold moves and a clearly articulated Vision for company people from the grassroots up to hear the clarion call of a new day.

Perhaps Gerstner's delay was the caution of a non-technical professional manager who found himself trying to lead a high-tech company; perhaps it seemed the appropriately judicious course.

And of course it's always easy to second guess. Judging by the stock price, Gerstner appears in the Summer of '95 to have Big Blue back on track, but from the perspective of hind-sight, it seems clear the company needed strong, positive actions much earlier in the day.

CEOs must make bold moves. That's the responsibility of a CEO, and the responsibility, within appropriate limits, of the leader of any organization. It's what defines the leadership role.

In a transformation situation, you've got to be bold because you need to get everyone's attention that you're on a new journey. This is not the time for timid plans.

Think it through carefully, but then go out and be daring. You must give people targets worth stretching for, and you cannot elicit the passions of men's hearts with indecisive goals. You must set initial vectors boldly. That's a key lesson to keep in mind: Being timid at the outset will only hurt you.

If you're planning a car trip from Silicon Valley to Los Angeles, would you plan it down to the smallest detail, would you worry about every traffic light, every right turn or left turn, every pothole? Of course not. Instead, you'd make sure your vehicle was ready, the gas tank was filled, you had whatever maps you thought you'd need, and you knew your destination clearly. It would be a waste of time to plan every move, even if you could, because by the time you finished planning, something would have changed and your plan would already be out of date.

What you need to work out in detail are the *initial* steps. You need to develop your driving skills. You need the ability to maneuver out of your parking space, and you better know how to get to the freeway and which direction to go when you get there.

The same thing holds true for business planning, and, in spades, for planning a transformation. The Vision tells everyone what the destination is, and the approximate roadmap you need to get there. That and the first few steps are what I mean by the initial vector.

But the Vision is not the place for specifics. Beyond the Vision, you also need some measurable goals that provide this initial vector.

A lot of people are guilty of over-planning. That's always a mistake, but when you're launching a new operation or transforming a laggard one, it's a mistake you can't afford.

There's a limited period of time when you can still seize everyone's attention and get them focused, put the past behind them, and get them ready to do something new.

At the first off-site with the National Semiconductor executives, I called for a return on equity of 20 percent—which is what the best companies in the industry achieve.

National's ROE was then a large negative number, and you could hear stifled, embarrassed laughter from around the room because what I was asking for seemed so unrealistic and un-achievable. That was my bold vector, subtle as a two-by-four. It was designed to get their attention, and it did.

The company's gross profit was then 24 percent. Another target I set called for a gross profit of 40 percent; three years into the effort, we had already achieved 43 percent. The difference represents some half-billion dollars annually.

These goals seemed unrealistic. But we wanted people to know they were being challenged to make a significant stretch.

If you call for a 10 percent improvement, your employees will just work harder. They'll produce the 10 percent for you, and in normal times that may be worth celebrating. But in a transformation, 10 percent does not do it; you must make demands that require a paradigm change.

At National we changed the entire corporate accounting system, set seemingly impossible financial targets, revamped the organizational structure (twice) and redefined some of our key ideas about what business we are in. And those are only a few of the "bold actions" we undertook.

Getting National's work force to believe in the possibility has been a cornerstone in the transformation.

People may snicker when you set bold goals. Ignore the doubts and make your expectations stick. The impossible only becomes possible when you believe enough to make others believe. There is a parallel with inventors here: many inventions are born out of the work of people who do things others thought couldn't be done.

COMMUNICATE WISELY AND WIDELY

To gain momentum in a transformation requires two things. In addition to taking those brave initial steps, you also need to communicate the transformation messages to all employees. Everyone from the executive suite to the plant floor must understand the direction that the company is trying to move, and must understand that achieving these changes is essential to success.

They must be convinced you're on a new journey. The changes at National sent a red alert.

But was everyone getting the right messages? Good communication is a fundamental essential, and not the old Telephone Game—the CEO tells the VPs, who tell the GMs, who pass it along to... If the message ever gets all the way to the backbone of your work force, it's sure to be watered down, garbled, and maybe even alarming.

It's surprising to find any company today that does not have effective channels for talking to its employees. Within two months after I arrived at National, we had hired Mark Levine, who is now director of employee communications, and he had begun to bring on a small staff.

Mark now oversees a full-scale internal communications program using all the familiar channels: a bimonthly journal, *InterNational News*, a video magazine, and communications meetings... to name just a few. These media trumpet our wins, sound an alarm about our problems, transmit executive messages, and advise our workforce about the change process. We congratulated our employees in print over establishing a foothold and major business contract in mainland China, and built their pride with a photo of Boeing's new 777 jetliner, featuring callouts showing all the places where the plane uses a National Semiconductor product. To encourage risk-taking, we did a humorous piece for the video magazine, that climaxed with senior vice president Charlie Carinalli diving off the high board into a swimming pool fully dressed for the business day in suit and tie.

I hold a company's managers absolutely responsible for ensuring that the important messages are accurately relayed to their people. I felt some early push-back, in the form of pleas for help from managers who didn't think they were equipped to stand up and explain the programs and messages. Levine tackled the challenge I presented by developing a series of materials called "Managers in Brief," which includes all the tools for the management communications sessions. As an aid in communicating the Vision, he and his group also developed what we call "a meeting in a box," with complete materials for managers to present the Vision to their employees—a suggested meeting agenda, a set of color overheads, a bullet-point script with suggested opening and close, and a list of questions to expect and the answers.

SETTING THE PATTERN OF BEHAVIOR

Organizations reflect the attitudes of the leader. In a shop where all the clerks are rude, you can expect to find they're taking the cue from the rudeness of their manager. If the leader of a company wants certain attitudes, or beliefs, or behaviors, to be universally subscribed to by the employees, the leader needs to make it clear that he or she is truly committed. This is another example of the current catch-phrase "walking the talk," but it's been true for a lot longer than the phrase has been around, and strong leaders have always known this intuitively. You need to exhibit the behaviors you want your employees to exhibit. And in a large organization, you'd better find some ways to let them know that it's more than just "Do as I say."

As you've already gathered, I believe in using off-site get-togethers as a way of building bonds and setting guidelines with your direct reports. It's easy to say that off-campus meetings are too expensive and too disruptive, but those are usually excuses. By sharing ideas on a more informal level, in a setting away from the daily pressures, you build long-term supporting structures. Besides, they can usually be done without spending a lot of money.

I'm also a great believer in the coffee klatch, a technique I've been using for more than a decade. At National, about once a month (more often, when I can) I sit down with a dozen people chosen at random from every corner of the company, but mostly from the grass roots, who gather with me in the same conference room where our Board of Directors' meetings are held. Each of the people is expected to ask at least one question, and even the timid people who came expecting just to listen usually find themselves engaged and participating before the session is over.

These people who would never see their CEO close up get a chance to discover that I am personally committed to the programs and changes they've been hearing about—many of which *originated* with me. But it's also a reality check because it's a chance for me to find out whether the programs and messages are really getting out. The people who attend aren't "on the line"—they're not being judged or graded. But the experience often brings to my attention issues I wasn't aware of. When I talk to the managers, I then have this additional perspective on the topics I need to raise with them.

The communications meeting is a widely-used technique for keeping employees informed. At National, we look on our company-wide quarterly communications meetings as a vehicle for publicizing the change efforts, and providing a forum for employees to learn about the different parts of our business from the managers themselves. But probably its most valuable function by far comes from the way we use this channel to personalize the news of National's wins. I share with the live audience, and with those watching on closed-circuit television around the world, stories of large new contracts, technology break-throughs, news of how well we're doing in sales, profits and stock price, praise from the financial community, and so on. While I try to keep these meetings balanced between successes (for the confidence they produce) and disappointments (as another way of pointing at the targets we need to continue focusing on), I probably err on the side of successes because of my natural optimism.

Is your communications effort as effective as you'd like to think? It's easy to lie to yourself, which is why I'm a firm believer in measuring employee attitudes. I have regular employee surveys done by an outside research firm, the first one only months after taking over. I won't pretend that the results have always gladdened my heart, but we don't do them for a pat on the back, we do them for a reality check. The results go to an Action Team made up of grass-roots employees; we created these teams as needed, at Corporate and at each National site.

The most recent survey was responded to by 15,000 employees, who answered a set of nearly 100 questions in categories such as Competitive Edge, NonStop Quality, Safety and Working Conditions, Customer Delight, Valuing Diversity, Empowerment, Employee Well-Being and Morale, and Rewards and Advancement. We rated highly on items such as "You are proud to be a National employee," "You have a clear idea who your internal customers are," and "You have a clear idea of what 'Customer Delight' means in your organization." But, as is true every year, the survey results clearly showed areas that employees rate on the other end of the spectrum. Aside from the topics of pay and communication—which in my experience are always on the list, no matter how good a job you're doing—we were alerted to the need for focusing attention on fairness in performance evaluations, recognition for doing a good job and for achievement, and managerial assistance in developing an employee's personal career development plan.

The driving principle behind the entire communications effort is this: make sure there are no enclaves anywhere in the company that are not getting the message.

BUILD PRIDE WITH NEAR-TERM WINS

When an organization is not functioning well, you focus on the problems. That's the natural temptation, difficult to resist.

But no business ever got fixed by focusing just on the problems. You cannot save your way into success.

If the coach of the track team started a new season by setting the bar at six feet, his high-jumpers would quickly become discouraged at their failures; they might keep on trying, but their expectations would be low and their hearts wouldn't be in it. Instead, the coach begins by setting the bar low, and raises it as his high-jumpers gain confidence and capability. The taste of success gives them encouragement to keep trying the progressively more difficult goals.

In the same way, you're going to build pride and the sense of accomplishment by creating a series of near-term wins.

There is a technique in the semiconductor industry and elsewhere called "high-low analysis." The yield analysis people take the worst wafer that's coming off the line, and the best, and then try to figure out what the differences are. The outcome of this Agatha Christie sleuthing might turn out to be a mis-aligned second-level mask or a design-rule violation; whatever the technical answer, they seek improvement by focusing simultaneously on what's wrong and what's right.

It's exactly that course you need to follow—looking at what you're doing right simultaneously with what you're doing wrong. And the work force needs to hear loud, clear, and *early* what successes the company is having on its new goals.

At National we set a number of initial short-term goals, targeted to be achieved by the following June, at the end of the fiscal year. Leading the list was the goal of achieving a record $2 billion in annual revenues, which would represent nearly a 20 percent jump in a single year.

By the time we closed the books on my second fiscal year at National, we were, indeed, a $2 billion company. The business was already healthier.

National Semiconductor employees around the world had given themselves a cause for confidence that their company was truly on the road to better times. They had the sense that they had achieved an accomplishment to be proud of.

I sent out a call for celebrations at every National plant and facility around the world. Corporate threw a day-long picnic featuring international entertainment and food. In Portland, Maine, employees went on an all-day Atlantic cruise. The Singapore plant sponsored a day-long beach party. The Penang, Malaysia, plants put on three separate parties so that employees from all of the round-the-clock shifts could take part. Germany celebrated with a picnic featuring employee boat races. It was no surprise that the Texas operation put on a lavish barbecue... but the Israeli plant also chose to celebrate the same way—with a Texas-style barbecue.

You need to provide the satisfaction of early success as proof that change is under way and is working, and as an enticement that will encourage your people to keep striving for continually greater achievements.

WRAP UP

"Setting the initial vector" means you have to do more than just adopt a new course. At the outset, you cannot afford to be timid; you must take bold, decisive actions to set transformation in motion.

You will settle on a very bold longer-term goal—typically two or three years. But at the start you must set and achieve shorter-term, more modest wins to build momentum. And then you make certain everybody knows about the successes. You do this by communicating to every level—celebrating the wins, and building pride with them.

You hold up the promise of being able, in time, to clear the six-foot high-jump.

END NOTES

1. As quoted in *Rockwell International Semiconductor Products Division*, Emory Business School case study OM88-101, prepared by Harbridge House, Inc., 1988.

4

Valuing What Your Customers Value

"You need to determine, reliably and specifically, what your customers value, and make that the basis of creating your products and services. Some of the answers may surprise you."

Representatives from Digital Equipment Corp. (DEC) came in to the Rockwell Semiconductor Products Division (SPD) one day, looking for a way to provide their customers with modems that would allow sending data and files from one computer to another at a remote location, over telephone lines. Gil knew what the DEC people wanted, but did not anticipate the transaction would turn into a learning experience that would shape his views on listening to the customer.

The company had already developed products that provided the electronic smarts for fax modems, and was successfully selling them. Unlike today's modems, though, these early units only supported the sending and receiving of faxes; a separate type—the data modem—was required for handling computer data.

The division hadn't been making data modems when Gil arrived, but had the capability, and he had pushed it through development. The new product presented three marketing options: the company could sell the chip sets themselves; or SPD could mount the chips onto boards that included all the supporting electronics; or they could easily take the process to the final step and develop an entire box with everything, ready to "plug and play."

But the product-line director and manufacturing head decided SPD should only sell the product in board configuration. This decision appeared sound, since it was based on the company's experience (and the experience of the industry) with fax modems. A modem is largely an analog device, meaning that it needs to handle nondiscrete measurements, such as translating a particular shade of gray from a drawing into so many millivolts.

For computer engineers, the familiar domain is digital—the 0s and 1s that the computer understands, and the world of exact numbers represented by a string of digits. When computer designers try to tackle modem design, they quickly find themselves in a forest of subtleties. Gil's people, on the other hand, had world-class experience and knowledge with modems, and could guarantee that their board modems would work in any fax machine, and meet the telecommunications requirements of every country. They couldn't make the same guarantee if they sold the chips, leaving other, less experienced engineers to figure out how to design them onto a board.

But the data modems of the day weren't nearly so complex, and, along with a more advanced chip, left room for a company to design their own modem boards around the SPD data-modem chip sets. Gil pointed out this distinction, which would make it possible for the division to sell data modems as chip sets, boards, or boxes.

The team from DEC was interested in providing their computer purchasers with a data modem carrying the DEC label, and were looking for a supplier. They listened to the presentations—which stressed the virtues and advantages of boards, and glossed over the other choices. At the end of the day, the DEC representatives said, "We know how to make boards and build boxes, so just sell us the chip sets and we'll do the rest."

Gil had to step in and overrule his own people, who were ready to say No. The company sold DEC some samples to use in designing their new product.

In less than six months, DEC was back. The design work was coming along pretty well, they claimed, but, "We have a lot of other things to work on, so we thought perhaps we'd buy the modems already on the board, and go wrap a box around it."

Three months later, they were back again. "Didn't you say you could do a whole box modem?" In the end, they signed a contract under which SPD would develop a box modem to meet their needs, put a DEC logo on it, and also develop the users' manual, product spec sheet, and the rest of the supporting materials.

Instead of buying $10 chip sets, in the end DEC bought $150 boxes, yielding a much higher profit contribution and much more satisfying customer-supplier relationship. If they had been stonewalled at the outset and offered only boards, the DEC people would simply have gone elsewhere.

The real value—or to use the current jargon, the "value proposition"—turned out to be different than anyone from either company had at first thought.

And, incidentally, SPD became the world's largest producer of data modems and integrated fax/data modems serving the emerging data communications market for personal computers.

Gil has never forgotten this lesson. He says, "You have to be prepared to do what your customer wants. If he wants a bag of parts that can be assembled like a HeathKit, sell him that. If he wants a fully finished box with his name on it, sell it to him—not just because you'll make more money, but out of a fundamental belief in letting the customer say what's important to him."

But the path to understanding customer values is filled with pot-holes.

In the early 1980s, the world of semiconductors was undergoing a change, shifting from a process technology called NMOS (short for N-channel Metal Oxide Semiconductor) to a technology called CMOS (Complementary Metal Oxide Semiconductor), which appealed to customers as representing a higher value because it uses much less power. But CMOS poses a much greater challenge in manufacturing.

Marketing came to Gil one day to report a groundswell of customers wanting CMOS chips in place of the NMOS products they had been buying.

Manufacturing was adamant—surely customers couldn't need ***every*** *chip in CMOS. Anyone who knew the technology could see that NMOS would serve perfectly well for a lot of the needs.*

Serious backgammon players know it's fairly easy to learn the fundamentals of strategy that dictate the best move; the problem is that, in any difficult situation, more than one strategy rule applies, and the challenge lies in deciding among them.

The same thing happens in business. Marketing was charging down-field toward that essential goal of satisfying the customer. To manufacturing, it looked like this demand took no account of the enormous cost and effort required; and because CMOS involved a more complex process, the production yields would be lower, which would make manufacturing look worse—additional reasons for foot-dragging.

This marketing/manufacturing conflict was carried up to the heads of the two departments, and developed into a clash that erupted into a shouting match at the weekly staff meeting.

Like the backgammon player, each was following a valid principle. But each had a faulty definition of success—to one it was customer satisfaction, and to the other, holding down product cost.

Gil had already set some ambitious goals to begin the migration to CMOS. Despite customer pressures, he knew it would have to be achieved through a timed, orderly transition.

Meanwhile some spadework revealed what should have been uncovered from the beginning—that customers were eager for CMOS versions for certain specific uses, but were quite satisfied with the performance of the NMOS chips for most of their applications.

Customers did not expect the world of semiconductors to change overnight; they were willing to accept a paced changeover as the entire industry abandoned NMOS and moved to CMOS.

A clear understanding of the customers' value proposition can save disastrous decisions and point to an appropriate course.

BUSINESS IS A VALUE DELIVERY SYSTEM

When business lecturers and writers look for an example of a company that understood its value proposition from the outset, Federal Express is usually near the top of the list. Fred Smith organized the company in 1971 and spent two years researching and planning. When FedEx started operations in 1973, it was against the wisdom of the experts and bucking the conventions of the industry.

It's become part of the lore of American business that when Smith had first presented the FedEx premise in a paper while a student in a Yale management course, the professor ridiculed the notion with the comment, "The concept is interesting and well-formed, but in order to earn better than a C, the idea must be feasible."

The post office at that time would carry your one-ounce letter anywhere in the country for 8¢, and this 8¢ entitled you to first class, airmail service. But how about a contract, legal paper, purchase order, or payment you wanted to have delivered the next day?

Fred Smith made a critical decision at the outset: other attempts to provide improved delivery service all depended on the commercial airlines, which left them able to offer nothing better than the *likelihood* of next-day delivery. Smith understood that to succeed in offering a service that would be perceived as a high customer value, he would simply not be able to depend on the airlines. He would have to be in control of when the airplanes departed, where they would land en route, and so on.

Smith proposed to deliver your document of up to about thirty pages, or a small parcel, at a charge of $13—more than 100 times the cost of sending it by U.S. mail. But he also offered virtually 100 percent certainty that it would be delivered the very next day.

Looking back, it's easy to see why people were skeptical. Eight cents versus $13. And to provide the service on a large scale would demand hundreds of airplanes, a vast fleet of trucks (the FedEx fleet is today among the world's largest), a gigantic processing plant, and one of the largest information systems imaginable. Altogether, it simply sounded preposterous. Yet Fred Smith had understood the value to customers better than the experts had.

More recently, in another example of correctly perceiving customer value, Federal Express spent $25 million to provide instantaneous communications with every driver in their fleet. Why? To give customers a unique level of assurance—"He's just turning onto Elm Avenue, and will deliver your item in ten minutes." For that confidence level, many customers will remain loyal to FedEx even though competitors offer lower prices.

Once you begin to think of your organization as a value delivery system, you begin to have a new perspective on what your goals need to be.

Eventually you'll come to be convinced that you need to be working, not from the inside out, but from the outside in—not providing what you think will be effective or what your product people imagine the market will want, but instead acting on what the voices outside, the customers, tell you.

WHAT YOUR CUSTOMERS VALUE

Customers buy from you because, like Federal Express, you're providing them not just a product or service, but something of value—a product that works, a service that saves them money, on-time delivery, readable documentation, information provided in a form that's immediately usable, and so on.

The business gurus advise you to "Ask the customer and do what he says"—but what's really required goes far beyond anything so simplistic, sometimes even beyond what the customer himself understands.

You need to determine, reliably and specifically, what your customers value, and make that the basis of creating your products and services. When you take the trouble of finding out what your customers really want, some of the answers may surprise you.

For organizations with internal customers, learning what they value should be a straight-forward task—made easier when you're willing to listen.

My management style is very different from my predecessor Charlie Sporck, who kept his door open to everybody but seemed to make all decisions himself. I did not expect it would be a smooth overnight transition away from a culture that had evolved over the twenty-four years of Charlie's reign, but I wasn't ready for the degree of resistance.

In my early days at National, while I was inevitably struggling to attach names to a lot of new faces, my administrative assistant Bonnie Murphy passed along a request I had made: "Gil wants a photo chart of the people who will be attending the meeting, with their names and titles." She was told, "We don't do it that way at National."

Looking back on the transitional period, Bonnie remembers encountering that attitude often. "I came to hate that phrase," she says.

People who try to stonewall their CEO are going to be even worse listeners when talking to the customer. But not listening well is a problem that, to paraphrase, afflicts some of us all the time, and all of us some of the time.

You need to create an environment in which people are open to listening and willing to probe, an environment in which they're dedicated to discovering what your customers really care about, and then will move forward with determination to deliver those things.

Several years ago, Westinghouse decided on a plan to increase its share of the refrigerator market in Mexico, against competitor GE. According to an executive who was one of the participants, the plan was to build an $80 million plant that would turn out refrigerators with significantly higher efficiency, and so prove less costly over the long run on the purchaser's electric bill.

No one, however, had done any market research until the last minute. It turned out that the market was indeed huge—only 40 percent of Mexican households owned a refrigerator.

But the rest of the research news wasn't as good. Most of the families that owned a refrigerator didn't have them in the kitchen, but in the dining room or living room. Why? Because the machine had been purchased as a status symbol. And most of them weren't even plugged in.

GE, meanwhile, had a marketing advantage—but one that was purely accidental. They sold their refrigerators covered in brown wrapping paper; in many homes, the paper was seen as a desirable addition, since it changed the white box into something that matched better with the appearance of the other household furniture.

Westinghouse canceled the plans for the new plant.

While admittedly this is an extreme example, the story nonetheless offers a memorable lesson in the need to understand what your customer values.

In my early months at National, I was applying pressure to find areas that could start producing income quickly. Steve Hamilton, Director of Marketing for the Memory Products division, who was one of the people feeling the pressure, remembers that "we had to find ways to do some creative things with very little." In the previous year, the division had lost $20 million, which was close to a record. They had to find ways of improving, but without spending a lot of money to do it. The notion of finding out what our customers would value offered one viable approach.

At an industry meeting, Steve and other Memory Products people bit the bullet—they approached one of the Motorola representatives present and simply asked him, "If you could have three wishes about memory chips, what would they be?"

In fact, the man had three good answers. Steve suggested that National might be able to help and asked if Motorola wanted to send an applications engineer in for a couple of days. The response was enthusiastic.

Out of the visit that followed soon after came a very good, close working relationship with the microprocessor division at Motorola, which led to a National program aimed at developing lower-voltage memory chips for their cellular telephones.

The first few times Hamilton went out to talk with other customers about the new 2.7 volt chips that the division was working on, Hamilton says they "looked at us with really blank stares, as if we were absolutely crazy."

Despite those initial reactions, National "hit the market for low-voltage E^2PROM memory chips right on the nose." The new chips mean that the user of a cellular phone or an electronic hand-held notebook may get as much as eighteen hours on a battery charge, instead of two hours or so.

In this case, National hadn't just asked customers what was on their Wish List, what would give them a competitive advantage, or what would make their life easier. The Memory division people had done all that, and then gone beyond to recognize a need the customers themselves were not yet aware of. In doing so, we created a new product category.

This product has lead to a number of others, one of them a combined effort with sometime competitor Intel to create a "plug-and-play" unit to operate under the Windows 95™ software. Our problem at the moment of this writing is that the division is not certain whether it can produce the units in enough quantity to meet the initial demand: a million every four weeks!

In 1994 the Memory Products division earned a $14 million profit—representing a $34 million improvement. The difference was that they had come to a vastly better grasp of what their customers value.

With customers, as with a spouse, you're likely to be wrong unless you ask. Understanding what customers value is, of course, one of those tasks we never complete because the answers are always changing, as technology and market conditions change. The Memory division, like all the rest of National, is still hard at work on improving their understanding.

In most cases, you'll find it's not at all obvious where the value proposition lies. The challenge of deducing the right answers is one of the more important things a leader does.

THE PROCESS OF DETERMINING CUSTOMER VALUES

Developing a sense of what your customers value frequently turns out to be difficult to do, and difficult to teach others how to do—most of the time you don't know what to ask, and even when you do, the customer doesn't know the answer. It's a squirrel hunt where nobody knows what a squirrel looks like.

The way around this apparent dilemma lies in recognizing that what's called for is not a one-time visit or phone call, but a *process*. It requires a getting-to-know-you, a relationship building over time. It's a kind of mating dance.

You need to develop an understanding of the customer's application, and they need to understand your capabilities. You must locate the point of engagement that represents the greatest economic gain for both. If you're overly rigid, you'll never find this point.

Here's a first step in teaching yourself how to understand the customer's value proposition: begin by understanding your *own*. Within a few months of arriving at National, I asked each division head for a statement of his or her value proposition, in one sentence, twenty-five words or less—the chief value that division offered its customers over the competition.

My request produced a lot of hand wringing and sweating palms, but not much in the way of answers. They didn't know, and the truth is I did not expect them to, because no one had ever asked before. This is, I think, not a standard business-school technique, and not widely enough used. It took us a couple of solid working sessions, talking about the methods and approaches discussed in this and the previous chapter, before the ideas began to take root and produce results in the form of value propositions for each major organization.

But the search for specific, particularized value propositions has guided a number of our businesses to remarkable improvements.

YOUR OWN VALUE PROPOSITION

To create your own value proposition, begin by writing down a first pass of what you think your statement might be. Make it brief, though not necessarily limited to the twenty-five words of my original challenge.

Professor Lynn Phillips of the Stanford Business School, together with Atlanta-based consultant Michael Lanning, has laid some of the groundwork in this area. We have adapted one of Professor Phillips' suggestions to express what we are looking for here. National people are told their statement should answer this question—

> From the customer's perspective, what do you offer, and what benefits do your products/services (or one particular product/service) offer in each key target market segment, that will lead customers to buy from you rather than from the competition, and at what price premium, if any?

When you have a first draft, try it out—within your organization and with some customers. Test it against reality. As you find what's wrong with the premise, change it.

Your value proposition needs to be stated in terms of satisfying your customers. Understanding your *customer's* value proposition requires that you understand *his* customers, and down to the end-user. If you're selling to Compaq, you must understand the CompUSA chain that sells Compaq computers, and the Compaq buyers.

As a refinement of this concept, we later devised the term "value *advantage*" to make sure the statement reflects the competitive distinctions—the reasons a customer would want our product in preference to any choice of competitive products.

A NOTE TO THE ENTREPRENEUR

Virtually every successful entrepreneur I know is today engaged in something different than his or her original business plan describes. Keep examining your own value proposition, and make sure you keep it up to date with—and if possible ahead of—your customers' perceptions of their own needs.

To cite just one example, LSI Logic, a Silicon Valley company headed by my friend Wilf Corrigan, was funded for one thing, but found their customers wanted something else, which they then proceeded to provide. What venture capitalists usually fund is not the product, or even the concept, but the people.

WRAP UP

What you deliver to your customers is not products, it's value—which has to do with the nature of the relationship, the kind of trust you create, the support you provide, and your ability to be flexible.

We all prefer dealing with suppliers we know we can depend on, companies or departments that are in touch with our needs. In the opposite direction, when you've got the right value proposition, you're able to build such strong relationships with your customers that they will be reluctant to go to the competition.

What's more, they'll willingly pay you a premium. That's a relationship based on an understanding of the value proposition, and it's the kind of relationship you should aim to have with all your customers.

5

Uncovering the Factors that Define Success: The Six Critical Business Issues

"A differing definition of success is the most common cause of conflict between managers and subordinates."

At the first Executive Off-site, on day one, Gil walked into a conference room at the Chaminade Whitney Center in Santa Cruz, California, looking relaxed and in charge despite the tension in the air.

The forty assembled executives weren't likely to challenge him openly, nor to rise up in some kind of spontaneous rebellion. But they didn't know what to expect, and were unlikely to embrace with enthusiasm any edicts handed down from on high by a CEO they barely knew.

Gil was in his tutorial mode. He had expected to spend his life as a professor, and only entered industry to gain some experience. While he never got back to the campus, he thoroughly enjoys teaching. His personal style has evolved to leading more by mentoring than by managing,

Five feet ten, with a mass of Mediterranean curly black hair (he may be one of the few Fortune 500 CEOs in the country whose hair is only just beginning to turn gray), Gil most of the time seems to be operating at low intensity. Often he seems too laid back to fit the image of a CEO; then he asks a probing question, and you realize he's far ahead, coping with problems you didn't even realize you had raised.

When he finds a management or technical concept he thinks valuable, he stores it in memory, and may recall it for use years later. He generously shares his insights and ideas—not demanding that people follow him, but giving them room to grasp the concepts themselves.

More impressive even than his memory, his patience, and his skill in gaining support, may be Gil's analytical ability. Charlie Kovac, who worked under Gil at Rockwell and has been a friend and consultant ever since, remarks on how well prepared everyone gets before a meeting with him. "He asks questions that jump right to the heart of your topic. If you haven't thought it through all the way, he'll find it out in the first few moments," Kovac says.[1]

The National executives at Chaminade Whitney that day in 1991 hadn't yet had much chance to find out for themselves about their new leader. Their tensions were natural—it wasn't at all clear that even a proven transformation manager could keep their company from being sucked into the quicksand that already threatened it. And, worse, even if the company survived, nobody had promised they would keep their jobs.

Wearing a dark gray open-collar sport shirt and carrying his very thick set of overheads, Gil took ownership of the room. What the executives saw at that moment was a relaxed, pleasant-seeming man with a presentation under his arm. They couldn't yet know that his comfortable grin was a real expression of his feelings toward them and excitement at the positive plans he had begun formulating.

Gil handled the visuals as he spoke, changing overheads himself on two separate projectors. (His aides wish he would let someone else do this for him, but Gil prefers the freedom it brings—even if his audience does occasionally have to cock their heads to read a crooked visual.)

To bring into clear focus where the company was off the mark, he shared with the group a basic but detailed diagnostic analysis of National, and a comparison to other major companies in the semiconductor industry, using measures like Gross Profit Percentage, ROE, and Revenue Per Employee (see Chapters 16 & 17). This was an eye-opening look, revealing weaknesses in the company that the executives had not fully recognized.

Following this lengthy tutorial, Gil announced a work session, and the forty executives regrouped into eight teams. The assignment for each team was the same: draw up a list of the main problems, the obstacles to success, that were keeping National Semiconductor from becoming great.

Instead of lecturing on the problems as he saw them, Gil was relying on their knowledge, collective experience, and forthrightness. He wanted their criticisms—something most managers never get asked for.

Gil had begun the process of involving his top team in the transformation process—making them participants in change.

Typically, instead of telling, Gil was asking.

Two hours later the group had produced a list of some seventy-five problems. But they weren't through. Gil next presented them with a follow-up challenge: he wanted their list of actions or programs the company would need to carry out in order to solve the problems the group had identified. And he wanted the **shortest** *solutions list they could devise.*

What the National executives came up with was a list of some forty items. Gil was not happy with this result.

The company would be attempting to get its entire work force engaged in the transformation effort, and clearly you'll never capture people's enthusiasm if you throw forty new programs at them simultaneously.

Looking back, Gil remembers the mountain that needed to be scaled. "What do you do in the face of so daunting a list? But I'm a synthesist by nature. Give me a group of items, and I'll try to make sense of them, I'll try to organize them by finding the patterns. That's how we made something practical and do-able out of the original list of forty items."

Following the off-site, Gil continued tackling the challenge, getting the list down to about eighteen categories—still too many for people to remember and cope with—and finally, after a month and a half, down to a manageable list of just five items, later expanded with the addition of a sixth.

At last they had a group of efforts that could galvanize the company. The list would eventually become known as the Critical Business Issues—the "CBIs."

THE KNOWLEDGE SOURCE

Writers are often asked, Where do you get your ideas? (a question which, incidentally, they hate—I suppose because they assume anyone who would ask such a question could make no worthwhile use of the answer).

In the same way, I'm often asked how I know what to do when I've just walked in the door. The natural impulse on moving into a new job is to demonstrate quickly how good a grasp you have on the situation by instituting new programs and projects, and perhaps making some snap personnel decisions.

These actions are reckless at best, and perhaps suicidal. Loading a lot of changes and new programs onto the shoulders of your new work force is hazardous to the health of all involved. It's the scorched-earth policy of the business battlefield.

In a transformation mode, changes are obviously going to be necessary. The question is how you divine what's needed. The answer lies in harnessing the people who already understand the organization, its operation, and its problems.

They may or may not have the right answers, but they have the knowledge and certainly know many of the problems. What you need to do is ask the right questions.

The Critical Business Issues provided an umbrella for the programs that National needed to address if we were going to survive and succeed. Our strategic plans, and the operational elements of those plans, need to embrace these CBIs very strongly so the plans don't just describe where we'd like to be, but also describe how we're going to deal specifically with the issues in order to get there.

The list will be different for each company, depending on situation and circumstances. Here's the National Semiconductor list of our original five CBIs, followed by an explanation of each (more fully detailed in later chapters)—

1. Organizational Excellence
2. Operational Excellence
3. Strategic Positioning
4. Return on Investment in R&D
5. Financial Performance

ORGANIZATIONAL EXCELLENCE

Organizational Excellence deals with key elements in a pragmatic process to create an organization that is strong, dynamic, and powerfully effective. At the core of Organizational Excellence is a process that focuses on making employees highly effective.

At National, we achieve these goals through a participatory, team-based, consensus-management process. Under the Organizational Excellence heading, we gathered programs aimed at these goals—

Direction Setting and Visioning

The Vision, as previously discussed, makes sure employees have their sights fixed on "the future that could be."

Eliminating Obstacles

This is an organized process for removing the bureaucratic roadblocks that too often keep employees from succeeding. Related to this is providing mechanisms that legitimize an employee's ability to raise issues on organizational dysfunction: "It's okay to talk about our problems." (As examples of efforts in this category, see the description of Leading Change in Chapter 7 and elsewhere, and the "100/100 Program" in Chapter 9.)

Educating

For National, creating more effective training and education was another area that needed to be addressed as an intrinsic part of transformation. (For example, see the discussion of National Semiconductor University, in Chapter 9.)

At one of the early Board meetings soon after I took over, I presented the concepts of the Organizational Excellence approach. The question from the Board was, "How soon can you get it started?"

OPERATIONAL EXCELLENCE

Under this heading come the nuts and bolts of how we go about delivering value to customers worldwide—cycle time, quality, service, manufacturing yields, and the other measures that determine how sharp we are.

Notice that these first two CBIs deal with implementation—the ability to get things done, to do them well, to do them right the first time, and to do them quickly. The greatest strategic plan in the world isn't worth much if you can't implement it, and I consider these two absolutely essential for doing this.

The programs to boost our Operational Excellence include some that were already in existence, and others created to address specific shortcomings. The list ranges from the standard—such as Nonstop Quality and Computer Integrated Manufacturing—to National programs focused on problem areas, such as our "Every Call Counts" effort to improve what was a very shoddy telephone responsiveness and a source of frequent customer complaints.

When you can achieve a high level of success in delivering value to customers, the marketplace rewards you with superior gross profit performance and asset productivity.

This is a never-ending quest, and National Semiconductor is still trying to learn how to do it well.

STRATEGIC POSITIONING

Strategic Positioning deals with identifying and achieving a business model that will enable superior financial performance. It defines a place in the business universe where we can plant our flag and claim the ground as our own. At the same time (to switch metaphors), it's like the company's periscope—providing input on issues such as how our customers and competitors view us.

One leading component of this CBI is benchmarking, which enables us to track progress toward goals. Another is the various programs that help us improve our image—globally, nationally, and in the communities where we are located.

The last two items deal with some of the key financial methods for measuring how well we're doing.

RETURN ON INVESTMENT IN R&D

This is a subject covered in detail later on (Chapter 16). For the moment, suffice it to say that while "R&D ROI" isn't a number you'll find on any financial statement, these metrics collectively answer the question of whether we are getting the R&D "bang for our buck."

As a technology company, our life blood is new products, new ideas, and new ways of doing things. National's success or failure as a company depends on the ability to transform R&D dollars into effective products that build on our history.

When you find that others are able to provide a clear explanation of your programs, it's a sign that your ideas are penetrating through the walls of resistance, so I was pleased to read an interview in an issue of the company newspaper *InterNational News*, in which a National vice president, Walt Curtis, described the R&D ROI. He told the newsletter—

> This Critical Business Issue is built around National's ability to quickly transform new knowledge, products, and services into increased profit and return on capital. We have to protect our intellectual property, and invest in products that contain lots of our core competencies.

FINANCIAL PERFORMANCE

For any company that can successfully achieve the first four tough CBIs, this fifth one will take care of itself. If you really have the others under control, Financial Performance falls easily into place. But that doesn't mean you can ignore it.

In financial terms the management and continuous improvement of the first four CBIs should yield—

- Steadily rising productivity;
- Declining break-even as a percent of sales;
- Rising revenue per payroll dollar;
- Increasing gross profits; and,
- Improving return on net assets and return on equity.

During our initial transformation phase at National, I put three specific items at the heart of the Financial Performance CBI. These three were gross profits, asset management, and our progress in reaching break-even. Even after achieving break-even, we have continued to track this value. (The critical financial issues are discussed at length in Chapters 16 & 17.)

When it came time to draw a diagram that would help communicate the concept of the CBIs, we quickly recognized that the five were not equal in importance. And so the diagram took this form.

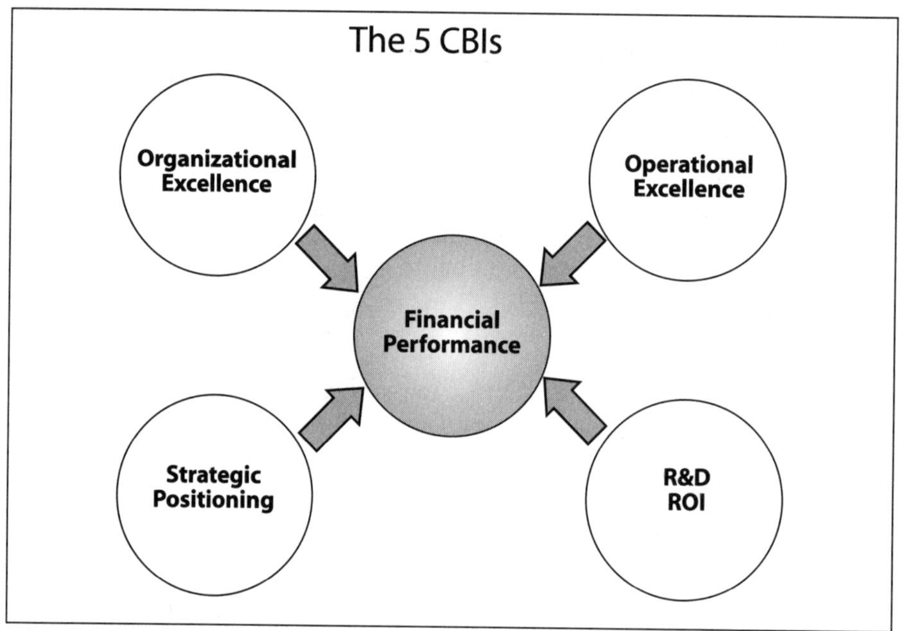

The 5 CBIs.

THE SIXTH CBI

When transforming an organization such as National, where most of the causes of poor performance were self-inflicted, the Critical Business Issues are at the outset almost always internally focused. Issues involving people, goals, organizational structures and so on are naturally and sensibly the place to start—so long as you remember the need to re-focus *outward* once the initial phase is showing progress. In too many situations the internal process drags on and on, weakening rather than strengthening the company; the transformation in these companies is blocked rather than encouraged.

It was a notable step in National's progress when about a year later we added CBI number six: Customer Delight.

The fundamental requirement of pleasing the customer is too basic and too familiar to dwell on at any greater length. But in order to drive home its fundamental importance and to remind all our employees that "customer satisfaction" isn't good enough, we substitute the phrase "customer delight."

Delight is much more than service, support, or satisfaction. Delight means helping our customers win in *their* marketplace, by delivering value through innovation and service.

(Our surveys tell us these are the two things our customers look for more than anything else: innovation, and service including quality. Are we providing them with service and new products that help them beat their competition, that help them come up with more compelling, more competitive new products of their own?)

When Customer Delight was added to the group of CBIs, we redrew the diagram to reflect this concept as a core element, the goal of all the others—

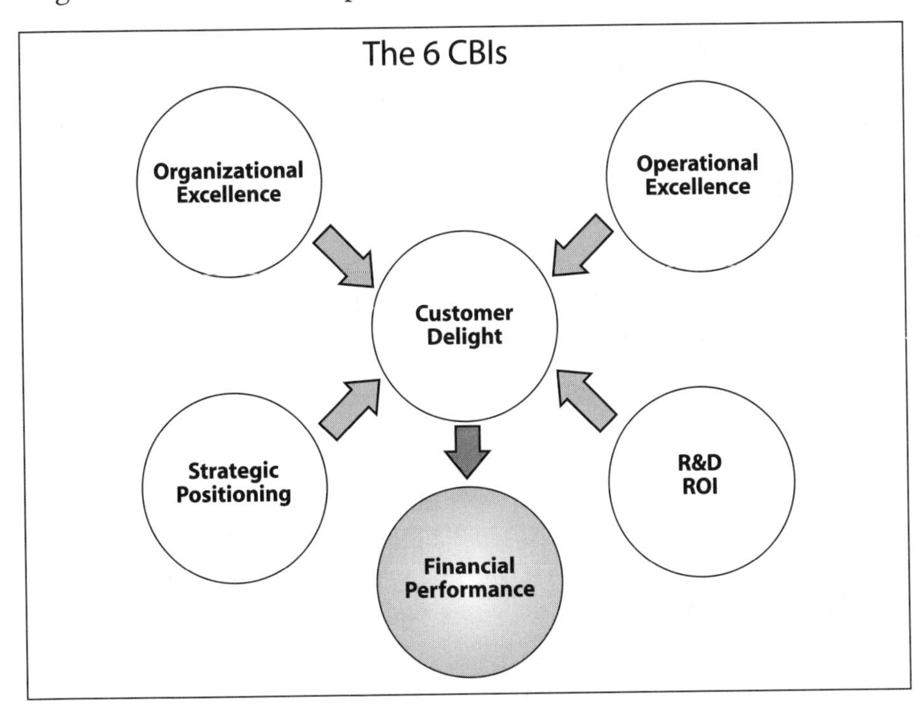

The 6 CBIs, with Customer Delight.

AN ASSIGNMENT FOR EVERY MANAGER

If CBIs are an effective approach at the corporate level, they are equally a good idea for every manager. At National, we encourage all our department heads and managers, when doing their goals for the year, to put at least one item under each CBI.

We say, "Take a look at these CBIs and ask yourself: What programs do I need in order to address each of these six issues?" And, "For each CBI, am I in position to support one of the corporate programs, or should I be creating a special program specific in my own area?"

In this way, the CBIs reach out from the corporate level to influence the focus of every manager.

EVOLVING

The story is told that just before the outbreak of World War II, the British discovered they were still paying a government salary for a civil servant to keep watch on a hilltop overlooking the Channel. The man then holding the post stood all day alongside a bell provided for his use, just as his predecessors had. His assignment was to sound the bell if he sighted Napoleon's fleet.

The tale may well be apocryphal, but it serves as a worthwhile reminder that even good programs may accomplish their purpose and become no longer necessary. So I'll reiterate a point made earlier—that one task for the CBI effort is a form of self-policing: to examine each program regularly and make certain it's still valuable and needed.

Some of the National's programs mentioned here will get the task done and be discontinued, and then we'll launch something new. These are processes that require regular policing.

The "Every Call Counts" effort is a good example. A time will come when our telephone support people will have absorbed the new attitudes and values so thoroughly that this particular program will no longer be needed, and we'll initiate a new effort to take them to a still higher rung of quality service.

In the initial efforts to pursue R&D ROI, we set out to create a better tracking system; we needed accurate metrics to determine our success. The tracking system is now in place, giving us the needed measures. The people who worked on this have now moved on to subsequent projects under this CBI—such as putting into place macro-measures of progress, and addressing core technical competencies.

CBIs FOR YOUR ORGANIZATION

I'd like you to look at the CBIs as building blocks for helping you and your entire organization move effectively toward the goal you've set as the definition of success. The CBIs are a powerful way to conquer the bane of every organization: how you get Mary Jane in Texas, Andrew in Pennsylvania, Ricardo in Tiajuana, and Uri in Tel Aviv, to have the same goals.

The bottom line on the subject of CBIs is this: You are currently running a number of what you hope are valuable programs—quality programs, financial controls, customer satisfaction, and so on. If your organization is like most, these are individual efforts, each making its own contribution to the whole, yet each pulling in its own direction. Although the intentions are powerful, the efforts are sometimes contradictory.

A program like the Critical Business Issues gathers individual talents and strengths together into a unified, coordinated effort. It challenges the leadership to make sound selections. It challenges you to determine which efforts demand early, concentrated focus, and which you can afford to put on the back burner.

You must involve the whole management team from the outset. They must be part of the process, part of the solution, and you should be taking maximum advantage of their help and insight. The particular method I used in creating the CBI program at National—getting input from the management team, and then synthesizing the long list myself—is not sacrosanct, and not even the best approach in all situations. But I was new to the National executives, and they weren't yet ready to work as a team at the level needed for this.

Whenever possible, let your management team be involved with the synthesizing, as well as with the problem identification.

WRAP UP

The CBIs provide a roadmap that coordinates management programs and involves all the people of the company. The aim is to set absolutely clear goals and to aid the management team in designing the appropriate strategies and tactical plans.

Sometimes, facing a plethora of business programs, it's difficult to tell which are short-term tools serving a temporary need, and which hold the power of lighting a fire under the entire organization. The CBIs are a firelighter. They have proven to be incredibly powerful—in aligning effort across the company, and in establishing a common vocabulary.

Although I had been using this approach since we evolved it during the transformation of the Semiconductor Products Division at Rockwell ten years ago, the method worked at National far better than I had any right to expect. The success came because it got people marching in the same direction, even from the very beginning.

END NOTES

1. Charlie Kovac interview, November 4, 1994.

6

On Selecting, Developing, and Keeping People

"Be tough on the issues and soft on the people.'

Gil Amelio's skills in working with people were originally forged in the crucible of necessity. In 1971, he was hired by the semiconductor division of Fairchild Camera and Instrument Corporation to continue research on his idea for an electronic technique of capturing images.

Arriving in the middle of an economic downturn, Gil found departments being closed and scientists being let go. The dark mood was palpable; it seemed an inauspicious time to begin a new job.

But the opposite would prove true. Gil would gain the knowledge about leading and managing people that has become the infrastructure to many of his practices.

In the face of receding income, Fairchild's director of Research and Development, Jim Early, was eager to find new programs that showed promise; the newcomer's project seemed to fit the description. Even though Gil had not had any significant management experience, Early tapped him to direct the research and also to manage the operation.

At age twenty-eight, Gil found himself running a department of seventy people. He recalls, "They were the créme de la créme of Fairchild, a bunch of fairly mature geniuses, creative people working at right angles to the mainstream—

which made them very difficult to manage. They were reminiscent of the early video-game developers—barely civilized." The description captures a sense of Gil's frustration, but also his respect and affection for the group.

In those days people didn't go back for an MBA to find solutions to their management problems, and in any case there wasn't time. Instead Gil haunted the library, where he discovered Peter Drucker's "Management—Tasks and Responsibilities." Although Gil had read other Drucker works, he found this one quite different; it offered what was then a novel idea: a view of management as a distinct role, rather than just a collection of additional activities that a business person had to contend with.

Business consultant Bob Miles describes Gil as "a sponge for ideas." Drucker's book became Gil's bible and could be considered a turning point in his life as a professional manager. He began to part company from other semiconductor-industry managers, who, he says, were locked into the mold of professional engineers and locked into an engineer's views; they "didn't develop a process of using appropriate management behavior for getting the results they wanted."

Reading one management book doesn't turn anyone into a manager. Gil was improving the handling of his Fairchild department, but not without a lot of continued fumbling and stumbling. Fortunately Jim Early was there to help him succeed. Gil remembers—

> *Of all the scientists I've ever known, Jim Early is up there with the very best. His ability to write clearly and speak precisely was highly prized; he set the high-water mark. He's the son of a reporter in upstate New York. He was very family oriented and having been raised that way myself, we had a bond on the non-technical side, as well.*
>
> *What I learned from him was not the hard side of management, but the soft side. Because he was caring, when you asked how to handle a situation, he'd respond with a solution that took into account the way people were feeling.*
>
> *He had an extraordinary understanding of the people dynamics, yet he could see this intellectually.*

Even at the time, Gil recognized the nature of the challenge: he "had to cope with people who were great technicians, but very difficult to manage." Today he often recalls what extraordinary training it was—with the pressure of seventy high-powered brains pushing at him, and the confidence and wisdom of Early giving him the support he needed to survive and succeed.

But Gil's success at Fairchild Camera was on more than one front. The research he was directing led to the creation of the first practical solid-state image sensor based on the principles of the charge-coupled device (CCD). In fact, more than half of Gil's (shared) sixteen patents embody the principles for designing and fabricating CCDs for image-sensor applications. The CCD image-sensor work done during this era would make possible the home video camera; some twenty-odd years later, it's still at the heart of these everyday products.

Typical of the Amelio evolution from engineer to leader, he remembers the big win at Fairchild in terms of the people, the team, and his rapidly developed skills as a manager.

OUT WITH THE OLD, IN WITH THE NEW?

Joe H. has just been hired as the new general manager of the Stafford Resort Hotel. Or brought in as the new head coach of the Dallas Cowboys. Or just been transferred from a small plant to become plant manager at one of his company's largest and most troubled facilities.

Over the previous three years or so, Joe has assembled an admirable, highly capable team of direct reports who are in part responsible for the success that has brought him this stellar new position. Which course is now most likely to assure his continued success—to bring along part or all of the proven team…or to wish the old group well and start laying groundwork with the people already in place at the new location?

It's frightening to leave behind loyal people you have worked with. They are people who know you, people you can depend on; they know your expectations and how you want things done. These are people who know how to give you space when you need it and know what information you need before making a decision. It's a risk to move forward without your trusted team to back you up.

But while I may not have the right answer for a football coach, in business my answer is always the same: start with the people who are already there.

What do the people in the new organization have that's so valuable? Their "institutional memory"—their knowledge and experience of the company—is so critical a commodity that you cannot afford to trash it.

No matter how skillful, a newly-arrived manager isn't tuned into the people or products, the suppliers or customers, the procedures, or the culture. What a help if the new manager can depend on a group of direct reports who know the strengths, the weaknesses, and where the skeletons are buried.

Perhaps even more important is the network of connections and favors that established employees have built. Bringing with you one or two people of your personal staff—a secretary, a staff assistant—is fine. But the line positions, the people who get the company's real work done, or who supervise those who do—that's a different story. Much of the work of a company is done not along any channel shown on the organization chart, but through the informal network of personal contacts—people one knows through the company bowling-team, training classroom, cafeteria introduction, parking lot neighbor... those one has dealt with through years of getting purchase orders written, personnel matters resolved, designs approved, new products launched and publicized, financial analyzes done, and all the rest.

Wipe out a layer of management and you disrupt this network. Conditions are almost guaranteed to turn much worse before they start getting better, and in a transformation situation, you can't afford the delay.

Lee Iacocca took over Chrysler and brought in some great people, but the company went another $2 billion in the hole before they started making money again.

So be warned; if you decide to bring in some of your own team, expect it to take many months for them to learn what they need in order to be productive or to support you effectively.

POINTING THE WAY

So you're going to inherit the management team left behind by your predecessor. But if the organization is in such need of transformation, they must be part of the problem—they must bear a large measure of responsibility for having allowed the situation to get so badly out of hand. How can you rely on them?

I offer this guideline:

> Don't attribute to malice that which you can attribute to ignorance.

Perhaps it's a form of xenophobia that we have a tendency to see as villainous those people who approach their work differently than we do—if they don't have the same work habits, if they don't follow the same management principles, if they don't relate to their people the same way, if they don't have the same view of what the problems are, then they must be evil. If one division is doing poorly compared to the others, the manager must be incompetent.

Those who do it my way are okay; everybody else is one of the "bad guys."

A judgmental approach like that courts disaster. As a rule of thumb, you can assume that 80 to 90 percent of your people are competent and well-intentioned.

On the human side of the equation, a large part of the story of my transformation of National lies here. I didn't see the executives, managers, or workforce as evil, but assumed they were able, dedicated people who didn't yet know what to do.

It's up to leadership to be clear on *what* people are to accomplish, not *how* they are to accomplish it. There is a time to address the "how-to"; initially, focus on the "what."

I recommend you start with confidence in your people, let them clearly know what you want, and then wait to see who comes through and who doesn't.

How long do you wait? I'll get to that.

MUSICAL OFFICES: NOT A GAME

While I'm firmly committed to working with the executives already in place, that doesn't mean they're chained to the same desks I found them at.

Of top managers who were at Rockwell when I arrived, almost all remained. Of those at National when I arrived, all but two are still here, four years later. But only one is still in the same job.

I focused on identifying the greatest strength of each executive, and in time moved each to a position based on that strength. The temptation is to focus on weaknesses, and send people to training, or replace them. That's a low percentage approach—it rarely works well.

Instead, focus on identifying strengths of your people. Any amateur can find the flaws and weaknesses; it's much more difficult to identify those qualities that are strong, and build on them.

Some years ago, an art student was taken by her elderly professor, a well-known artist, to the Philadelphia Museum of Art to see a particular Rembrandt portrait, a 1645 painting of an old man that people travel great distances to see. The art student stood in front of the painting and felt very knowledgeable as she pointed to the inadequacies and mistakes in this great work of art.

She found the figure's left hand painted amateurishly and noticed it had only four fingers! She pointed out to her professor how the cane in the old man's right hand was repainted at least three times in different positions, and she scoffed at the flaking paint.

Her teacher nodded and then patiently explained the brilliance of the way the head was painted, the skill and technique of the magnificent eyes. Those who really know painting are in awe of how, with just a few brush strokes in exactly the right colors, Rembrandt captured the joys and sadnesses of a lifetime in those eyes.

It takes knowledge to notice and value worthy qualities. The overall greatness of this Rembrandt portrait lies in what's right about it. That's what makes it worthy, what makes it a valuable addition to the museum's collection—a painting worthy of being called a masterpiece. With strengths so great, mistakes become unimportant.

Not just in art but in all things, it's all too easy to see what's wrong, much harder to focus on what's right.

MIDDLE MANAGERS: CHANGING TEAMS

So far I've been talking about the value of working with the management team you find in place. For the leader of a large organization, this principle applies to your direct reports, and the next level—the people who report to them.

How about the middle-manager structure? It would be nice if the same principle applied. It doesn't.

A transformation involves a culture change. The old culture will have self-selected a manager population that suited its values, but the new culture will require different values.

Some people will not be able to make the change, and inevitably you will have to gradually bring aboard new talent. In the best of situations, this will involve only a small proportion of the manager corps.

National wasn't so fortunate. A number of mid-level managers were particularly weak, and we found it necessary to bring in a lot of new blood. This is a painful process for the individuals who leave, a challenge to the new arrivals, and a strain on the organization. But often it can't be avoided.

IDENTIFYING STRENGTHS

As I've described earlier, from my arrival at National I began taking a read on each of the people reporting to me—mapping out a mental profile of their strengths and weaknesses. I allowed myself four months to gauge the capabilities of the top team, based on their grasp of the business, their management skills, and how they expressed themselves at meetings.

One other factor I consider especially important at higher management levels: Where do his or her solutions come from? Put a marketing person into a troubled area, and he'll come up with marketing solutions. I suspect some dentists of believing that if everyone on earth had good teeth, there would be no more war.

Most people in business, as in other areas of life, are narrowly focused, and look in their own area of strength for the answer to every problem; while this is to be expected at mid-manager levels, it can be disastrous for the division manager/general manager, and up. In top executives, I look for a broader perspective.

ASSESSING PEOPLE: THE WELCH DIAGRAM

Jack Welch will be surprised to find out that I've given his name to a principle I follow and teach.

In October, 1991, I attended a forum arranged by the publishers of *Fortune* magazine for CEOs of companies on the Fortune 500 list. As noted elsewhere, I long ago concluded that the most valuable transactions at conferences and the like are often those that take place in the corridors. Much of what I have learned in management, other than from direct experience, has been in situations like this, where I picked up something I could not have anticipated.

In a conversation with Jack Welch during a break at the forum, we exchanged ideas about what we were each agonizing over at the time. It turned out that an item leading the list for each of us was the same: what to do about senior managers who were sluggish in responding to the change messages.

Afterward, thinking over the conversation, I came up with a diagrammatic way of depicting the conclusion that we had both arrived at. Because of its origin, I have ever since referred to this as the "Welch Matrix."

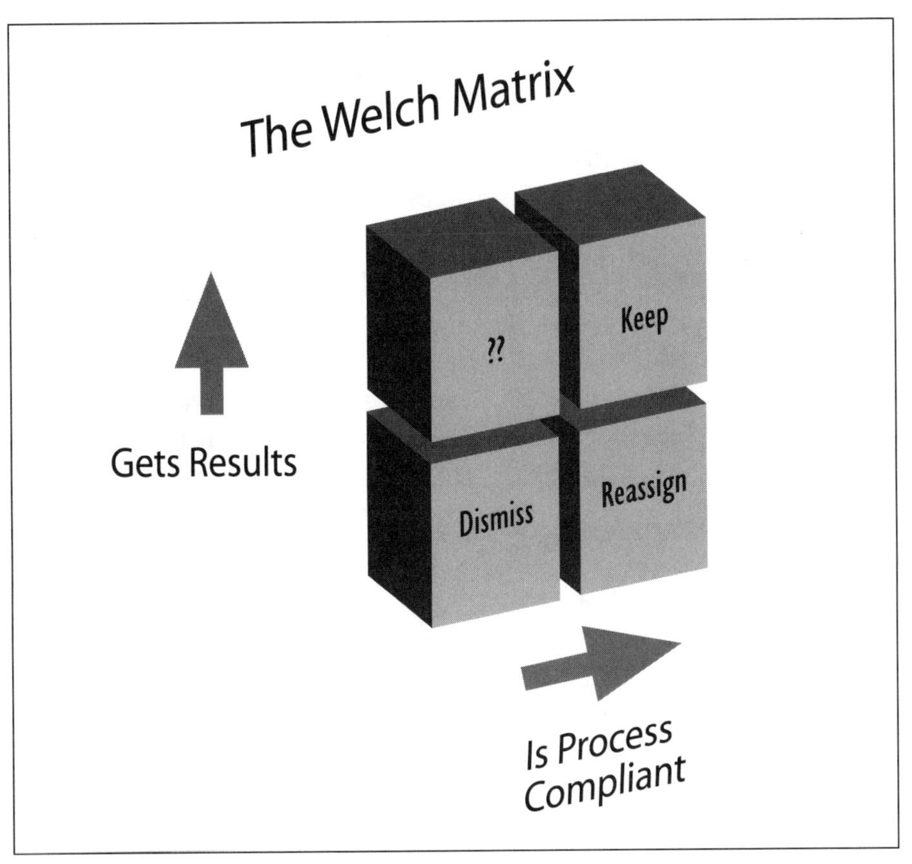

The Welch Matrix—Manager Characteristics.

The matrix is intended as an aid in arriving at a decision about each of four types of managers. It's based on evaluations of two criteria: how successful they are at getting results; and how compliant they are at following the character and culture of the organization; note that in a transformation, "process compliant" must be based on *today's* culture, not on what was acceptable in the past.

Managers who are getting results and are process compliant (upper right box) are the ideal. These are the "keepers"; you would clone them if you could.

Next best are the dedicated, industrious managers who work within the culture of the company, but are not getting results (lower right). These people are probably in the wrong jobs. Reassign them to positions that provide a better match for their strengths, talents, interests, and capabilities, and you'll likely move them to the keeper category—to your benefit and theirs.

Managers who are neither process compliant nor getting results (lower left) are failing the company. Satisfy yourself that it's not at root a failure of the *company*—that it stems from a lack of training, wrong job assignment, or similar. If there's no such cause, you need to let the person go.

These are easy categories, probably obvious. The dilemma comes with managers who fit in the upper left category—not process compliant but nonetheless achieving results. They may be unwilling to work within the culture of your company; they may be so awkward in handling people that their workers continually file complaints, ask for transfers, or quit. Whatever the situation, the challenge is: Do you tamper with success? Which is more important—the process, or the results? Is this a short term gain that will erode a long term goal? Decisions like this are critical to successful transformation.

You take over a department or a company, and one of your managers is achieving every productivity goal but appears to be a Neanderthal in dealing with people. What do you do?

The traditional wisdom is, if managers are getting results, leave them alone. "If it ain't broke, don't fix it." But Jack Welch and I had both reached a far different conclusion: "Fix it *before* it breaks."

If your processes are the right ones, then it's only a matter of time before managers in this category will fail; replacing them early will save you having to recover later on from the much larger problems. Rigidly resistant people who are not process compliant, even if they appear to be getting results, are people who are unlikely to grow and develop.

So the surprising conclusion of the Welch Matrix is that process compliance is more important than just getting results.

One caveat, however: Sometimes a manager who appears not to be process compliant is in fact pursuing some insights that are at the cutting edge—based on knowledge or techniques the rest of the company hasn't caught up with yet.

In conversations with someone like this, I strive to hear and understand their maverick perspective. It becomes a great moment in your own personal history when a maverick convinces you—when they convincingly talk their way out of the box.

Spend enough time to make the right decision. Before I fire someone whose performance demonstrates they belong in this category, I suggest a correction program, with a time table. I'm clear about what I'll be looking for to show that development is taking place; specific accomplishments are important in this discussion.

At the senior VP level, I'll allow a whole year for change to be demonstrated; with a supervisor or department manager, you will probably want to move more quickly. At the end of the allotted time, if you haven't seen significant change, you never will.

Three months after the original conversation with Jack Welch, the newspapers carried a story with headlines like "Welch Fires Four Sr. Managers." I knew exactly why; and now, so do you.

CHOOSING PEOPLE: ATTRIBUTES TO LOOK FOR

No one is right 100 percent of the time in choosing, retaining, or promoting people. But you can learn to greatly improve your batting average by focusing on the most important traits. Over the years, in decisions involving hundreds of individuals, I've evolved a set of yardsticks—qualities you should look for in candidates. Here they are—*in order of importance*:

Strong Personal/Ethical Qualities

> Admittedly not easy to evaluate in a 30- or 60-minute interview, but look for people who are fundamentally ethical, honest, clear-eyed, and forthright. With some candidates we tend to think, "He's so brilliant, I'll put up with this problems"; this is a prescription for disappointment or disaster.

Brightness

> This is not a question of education, but rather of native qualities. Admire natural curiosity, inquisitiveness, the ability to

catch on quickly. Look for wide interests—even (or perhaps *especially*) in an engineer or scientist.

Experience

This is the first thing most people look for—which may be one of the chief reasons so many get hired who don't work out. Of course experience is important, but in its place: it should not be the top item. Really bright people can learn what they need to get the job done.

Sense of Humor

Many will be surprised to find this item on the list, but the workplace is often a high-pressure crucible. The tension can be defused when someone in the group can bring a laugh or can laugh at themselves.

Ability to Work Hard

These days more than ever, prize a high energy level. The business environment demands a healthy stamina and enthusiasm. People with low energy—especially if they are complainers—can drag down the people around them, creating a situation in which transformation will not move in the direction you have in mind.

Intuition

Another item that's difficult to judge except when promoting from within, this was one of the toughest evolutions for me personally, and I acknowledged myself as a professional manager when I realized that I was using intuition in decision making—and without embarrassment or apology.

In the real world, you can never get all the data. Good managers learn to respect their gut instincts. In aviation, this kind of ability is sometimes referred to as "flying by the seat of your pants."

I've been successful in hiring and promoting when I have rigorously adhered to these six attributes, and in this prescribed order of priority.

ADDITIONAL PRINCIPLES

Here briefly are a few of my additional guidelines—

Hiring a general manager or above

The first thing I look for in a candidate for a high-level position is a quality mentioned earlier in this chapter—whether their solutions all come from a single category or area. Take a typical successful engineering manager, promote him to GM, and, unless he's already demonstrated balance and breadth, you've probably just invoked the Peter Principle.

Some people have the innate ability to approach a problem from more than one direction at the same time. Since I've never come upon a way of training people to think with multiple perspectives, I'm stuck with looking for people who already have this ability.

The Personality matrix

Be suspicious of the personality matrices that are widely used by HR and training departments, intended to help the manager or executive in evaluating capabilities of employees and candidates. I believe at least half of your judgment of an individual must be subjective. These forms which have you do numerical scores of people's qualities and abilities to three decimal places are, in my view, basically a waste of time.

Lateral moves

One of the ways to identify and develop talent is to put a lot of emphasis on lateral moves. I believe the only way you find out about your junior people is to see them in different settings.

When developing younger people, make it a practice to move them around to several different job environments early on—don't leave them in one job for more than two years. By the time they are in their early thirties, you'll have a good fix on their capabilities, and will be able to help shape their careers in ways that will most benefit them and the company. You'll also get a good reading on their ability to simultaneously juggle multiple viewpoints and handle them with balance.

The loner

I will not promote the brilliant worker who is a loner as fast as the solid but less bright one in the next cubicle who is a team player, who leverages by drawing on the assets around him, and who helps others by contributing to their success. Transformation demands people who can work in combined efforts.

WRAP UP

When brought in to head up an existing organization, count on working, at least initially, with the direct reports you find in place.

Start with the assumption that your people are competent and supportive, even when there seems to be evidence to the contrary. Give them a chance to climb on the band wagon. Remember: Never attribute to malice that which you can attribute to ignorance.

The lesson of the Welch Matrix is that process compliance is more important than just getting results.

In hiring and promoting, value strong personal/ethical qualities and brightness, over experience. Also give credit for a sense of humor, the ability to work hard, and intuition.

A lot of eager young people are ambitious for their own success; often I sense a time-table for success ticking away in their heads. If this description applies to you, try to refocus your ambition and your measurement of success. You are likely to enjoy greater success if you put ambition for your company, your department, and your team, ahead of your personal ambition.

In my view one of the greatest wastes of a nation's or company's resources is the people who never achieve their potential. I have always had a personal goal of finding ways to help as many people as possible rise to the maximum of their ability.

> *An ancient Arabian proverb says,*
>
> *He who knows not, and knows not he knows not—he is a fool. Shun him.*
>
> *He who knows not, and knows he knows not—he is simple. Teach him.*
>
> *He who knows, and knows not he knows—he is asleep. Wake him.*
>
> *He who knows, and knows he knows—he is wise. Follow him.*

★★★★★

SECTION 2

Creating Commitment: The Desire to Contribute

7

Institutionalizing Excellence: The Leading Change Program

"Your principal goal in designing a transformation program is to ensure no one will forget its central messages—that business as usual is no longer acceptable, and that every manager and supervisor has a role in bringing about change."

In the 1960s and 70s, the semiconductor industry was a component business. A company that wanted to build a new electronic product would call a supplier and order some chips called NAND gates, some others called NOR gates, perhaps a few accumulators, and then would assemble them, Erector™-set style, to create their product. The chip manufacturer didn't know if the components were going to be used in a calculator or an elevator control system, and didn't need to know.

National Semiconductor came of age in this environment; the company boomed, its whole culture molded around this way of doing business.

In the 1980s, the world began to change. The ability to put more and more transistors onto a chip meant the semiconductor manufacturers could create chips with more "smarts," and customers began creating their designs in a new way—designing products that would take advantage of the architecture which the chip-makers had already etched onto the silicon, or defining their own architecture and getting the chip companies to craft it onto the chip.

By the mid-80s, the new approach was firmly entrenched, and has been ever since. The architecture of most personal computers is defined by the microprocessor and, hence, by the chipmakers; it's only a slight stretch on reality to say that the computer is on the chip, and in terms of basic architecture, IBM and Compaq don't do much more than put a box around it, integrate software, and do some creative marketing.

As a result of this change in the chipmaker's role, the whole relationship between chipmakers and their customers has shifted dramatically. Chips are no longer selected in an arms-length negotiation based on price, delivery, and quality. Instead, the supplier's engineers work in close relationship with the customer's engineers to find the best solution for the end product.

National Semiconductor recognized this change at only one level: they could put more "stuff" on the chip. They responded by designing more complex chips and by looking for ways to make them cheaper and cheaper. Doing what seemed to be all the right things, they couldn't understand why sales were faltering.

What hadn't been changing fast enough was the culture—the collective assumptions that Peter Drucker refers to as your "theory of the business."[1]

Gil Amelio, at the time running a group of divisions in the telecommunications field for Rockwell, saw all this from the perspective of a customer, and concluded that Motorola and Intel understood the needs of their customers, while two of the other semiconductor companies "didn't get it." The two blind-folded companies were Texas Instrument, and National Semiconductor. Neither had recognized the need to build an intimate relationship with the customer. TI ultimately got the message and re-emerged as a first-class semiconductor supplier; National still hadn't deciphered the problem when Gil walked in the door.

There was an incident when one of Gil's Rockwell companies needed a large custom chip for a major product they were designing—a $2 million telephone switch. Gil recalls—

> *It was clear that National Semiconductor really didn't understand the idea of working intimately on the design of a critical item that could make or break the product. What they needed to do was to recognize that the decision the customer was making was based on the ultimate intellectual content of the semiconductor material, and therefore necessitated a much closer relationship.*

The world was changing, National wasn't; they went from being one of the stars of the 1970s to one of the dogs of the 80s—a classic example of what GE's

Jack Welch describes as "You drive change or change drives you."

When Gil arrived at National Semiconductor, the company still hadn't figured out why their business was disappearing. When National's executives heard Gil's explanation, the typical chagrined reaction was, "It's so obvious."

★★★★★

On a chilly Sunday afternoon in March, 1992, one of National's Total Quality Managers, Savitri Saldana, arrived at a beach-side complex in Monterey, California, to begin a week she had been anticipating. Saldana and twenty-five others had been selected for a course in a new program called Leading Change, a development opportunity which she understood was to be a kind of "mini-MBA" combined with a crash course in the transformation underway at the company.

Savitri found herself in a group that included vice presidents, general managers, and directors of National Semiconductor operations—an intentionally diverse group from facilities in Hong Kong, Singapore, Malaysia, the Philippines, Scotland, and Germany, as well as California, Texas, Utah, and Maine, and with a range of professional backgrounds that included engineering, sales, production, and management.

In a week packed with experiences she still talks about in a rush of enthusiasm, Saldana remembers a first intense shock, and then the resolve that followed—

> *I didn't realize just how much trouble National was in. I never knew that the company was on the edge of bankruptcy. That scared the heck out of me and then, almost at the same time, I said to myself, 'I'll do everything I personally can to change this situation.' Somehow it put a sense of passion in me.*

By this time, the company Vision had already been rendered into art form. It was a preliminary version, the class was told; they were invited to thrash out their ideas and arrive at suggested changes they could agree on as a group. The offer conjured up images of a suggestion box that's never going to be opened.

Not so—the next iteration of the Vision drawing contained some of their additions. Saldana says, "To this day, I can look at a certain spot on (the drawing), and say, 'I had input on this area.'"

One of the great attractions of Leading Change—one of the elements that made National managers clamor to be included—was the opportunity to spend time eye-to-eye with the new CEO.

She remembers, "It was wonderful when Gil got up at the blackboard," in his dual role as teacher and also as "Captain of the Pep Squad." Gil was intense about making sure that all change messages were fully understood. By personally delivering these messages and explaining their vital importance in his compelling style, he could insure both understanding and impact. Gil leveled with the participants about the current state of the company, its problems, what National was capable of becoming, the urgency of the transformation, and their role in making it happen. This was Gil Amelio at his best—speaking informally, without a script, about issues he feels deeply.

Saldana remembers her pride as she realized that National was being led by someone who was willing, not just to dictate, but to share, listen, and consider the ideas of others.

> *Gil was open and believable, and we were able to ask him anything and get an answer. He makes every person feel valuable and essential in the transformation. Gil says, 'Here's where we need to go, what do you think about it—let's create it together.'*

Saldana recognizes that although it's not how Gil wants to be thought of, the National employees "put him on a pedestal. That's because he puts us on a pedestal, too—not many leaders are strong enough to include their people in the planning."

WHY ARE WE STILL TALKING ABOUT CHANGE?

If half the articles on business management in the last twenty years have been on the subject of change (only a slight exaggeration), why do we still need to be talking about it?

Institutions don't change; only people change. When people change, they bring about change in their organization. In order to change the organization, you must find ways to change the people.

Change efforts don't work in most companies because they are instituted piecemeal—a new quality program here, a better customer-response program over there, a restructuring, a new Vision statement.

But the situation facing an organization in trouble demands something more fundamental: a coordinated, systematic effort that changes many things simultaneously. It's a theme you'll find repeated in these pages because it's one of the most basic cornerstones of successful transformation.

To make a profound and lasting difference to the organization, change needs to be part of a process. This is the essential nature of transformation—an on-going, focused, coordinated *process* of change.

MOTIVATION TO CHANGE: A PERSONAL STORY

It's the responsibility of an organization's leader/manager to perceive the factors that make change mandatory, devise the change process, and drive the changes through the organization—overriding the reluctance that will always be encountered.

The question is, how do you create an environment where change is more than just a topic everyone agrees on, but instead becomes a way of life, something people actually live by?

When I went to Bell Laboratories with the ink not dry on my Ph.D., it was the Mecca, probably the best industrial research institute in the world. At Murray Hill, where I worked, they had about 5,000 employees—1,000 Ph.D.s, and 4,000 others to support them, and the Lab was extraordinarily selective. For a professional position, you had to have a doctorate in one of the science or engineering disciplines even to be considered, and they hired one out of perhaps every 300 applicants.

I arrived in New Jersey full of confidence and a little pleased with myself. On my first day, after the usual morning of filling out forms, I was invited to join a few of the other young scientists for lunch in the Bell Lab cafeteria. We got our food and sat down—eight people gathered around a table.

It was all very pleasant until the introductions were over. After that, I didn't understand another word throughout the meal. Every subject they dealt with was over my head. I didn't understand the vocabulary, I didn't understand the equations, I wasn't even sure what areas of science they were talking about. *Everything* was over my head.

Walking back to my office, I was thoroughly depressed, unsure of myself, and wondering how in the world I would ever survive.

A terrible thing to happen? No—in fact, it was highly motivating. I wanted very much to succeed. I needed to be a credit to my family, my university, and my science. Anything short of success would have been humiliating. And so I proceeded to work my a__ off.

The environment at Bell Labs was highly competitive, and very stimulating. So much of the science that high-technology rests on today came out of Bell in those years.

It turned out to be the most stimulating time of my life, and an extraordinarily productive period that yielded several patents, and recognition through scientific papers and speeches at technical conferences.

I believe all this was due to the *stress* of the situation—a good kind of stress, a creative tension, that forced us to our highest levels of achievement.

CREATING AN ENVIRONMENT FOR CHANGE

This experience at Bell Labs would later lead me to an understanding that the very best and most creative organizations, those that are world class, are not *comfortable.*

I would eventually come to see that the most effective way to produce an environment of change, and create a productive and rewarding work experience, is through creating an environment similar to the one I found myself in at Bell.

Your chief job as a leader/manager is not to protect your people from the slings and arrows of the world, but to create the atmosphere of productive stress that drives change through the organization. You keep people running down a hill so steep that if it were one degree steeper, they would fall off the edge.

This is difficult to get to, and difficult to maintain, but it's what you need to be aiming for.

THE WRONG KIND OF STRESS

Just as there is good cholesterol and bad cholesterol, so there's good and bad stress.

Bad stress is when a manager creates aggravation by erecting obstacles and using bureaucracy to impede. Some well-intentioned managers turn out to be speed bumps in the parking lot of life.

Whenever a manager has the funds available but denies a well-justified request for new equipment, training, access to a database—anything that could enhance productivity or move toward meeting the goals—that manager is blocking the progress of the entire organization.

CONTINUOUS IMPROVEMENT AND QUANTUM CHANGE

You will always be seeking continuous improvement, but as I pointed out in an earlier chapter, sometimes—especially in a transformation situation, when ten percent improvement isn't good enough—you need to emphasize the more difficult goal of quantum change.

It's important that you let your people know which items on the change list are continuous, and which are quantum. I simply put a C or a Q in front of each item on my list.

OVERCOMING RESISTANCE

Cynthia Scott, a San Francisco-based psychologist who is the author of *Rekindling Commitment* among other works, focuses on the psychological aspects of business. She's one of the consultants I have turned to at National. On one occasion, coping with the challenges of transformation, I asked her for a model of how people change. She responded by drawing a diagram like this—

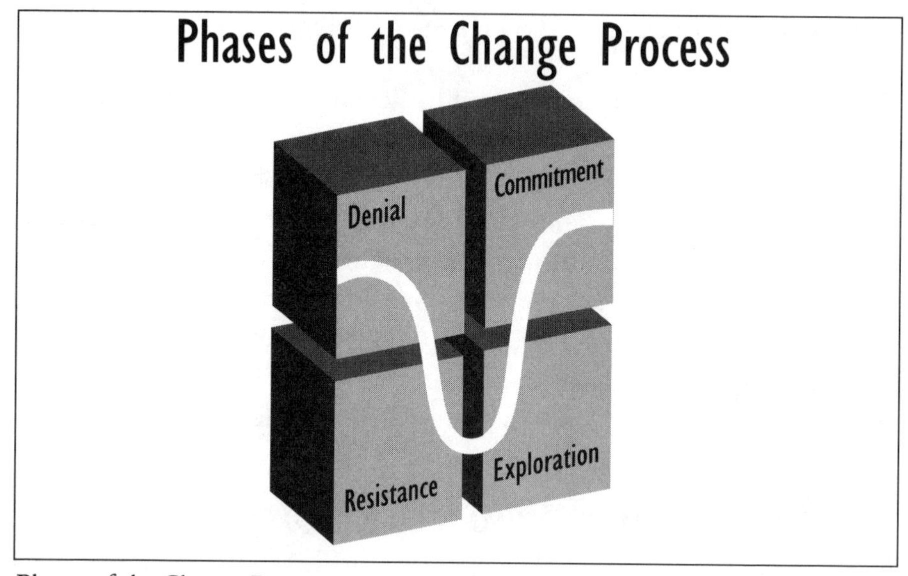

Phases of the Change Process.

A particular change, for a particular individual, will usually go through these steps, starting with "I don't want it" or "We don't need it" or "Around here we've always done it this other way."

Managers often hold the mistaken belief that all of their people are happy and comfortable, even in an environment of change. This is an Ozzie and Harriet version of reality.

To foster change, you need to expect and allow for the initial push-back of denial, and then move people through the steps of this process. If you try to force change without going through these steps, people will simply slide back. The process can't be rushed.

GATHERING DISCIPLES OF CHANGE

After taking over at National, I began to search for an effective approach to move the notion of being a transformation leader, down into the organization. I wanted to get "disciples" who would begin preaching the transformation religion on their own and spreading the word, so it would continue to replicate itself all the way to the grass roots.

My goal was, and is, to charge up everyone to become not just a student of transformation, but a teacher of it at the same time.

The fundamental premise of my answer was that people become committed when there's a very obvious self-interest. If a person feels good about coming to work, the work experience makes them feel good about themselves. People in that situation find it very easy to become committed and devote immense energy and effort.

So how do you help people feel good about themselves in their work situation? Examples of things we've been doing at National would include improving the communications, giving recognition (both covered in other chapters), and sprucing up the physical environment.

Whether office or plant, the physical facilities need to be clean and tasteful. Every professional should have a computer or access to one. My guideline is: Ask yourself what standard these people would set if they were creating an office at home, and provide a work area at that level.

You try to design a task, create a structure to make sure people's views are heard, and ensure that individuals can make a difference.

That's energizing. When people have the sense that "I can make a difference," then they can feel positive about their work.

It all revolves around this notion that everyone can make a contribution. Good leaders have known this all along.

THE PARADE CALLED LEADING CHANGE

The job of a manager is to manage change. How long would it take to determine what change was needed, and how would we actually go about making change happen?

By December '91, seven months after I had taken over, the underpinnings of transformation had been set in place. We had shaped a Vision, identified and begun to address the most urgent issues, and launched some innovative channels to communicate these efforts throughout the company. The top managers had gathered for three of the new quarterly off-sites; most of them had become supportive of the new approaches. The attitudes of the management team had in most cases evolved from willing to supportive. As we began to see progress with some of the uncomfortable steps that were needed to make the company viable once again, I could observe signs of growing enthusiasm.

When you initiate transformation, your direct reports are comparatively easy to reach. For people below that level, the sounds of progress are like a parade they can hear in the distance but cannot see—they're aware of cheers and shouts, fanfares and drums beats...but it's someone else's parade. They hear the noise but can't enjoy the spectacle. Someone else's parade isn't anything to get excited about.

You can talk about change, you can send memos, print articles, distribute videos—but by themselves they're mostly parade noise. Your people know something is happening, they may be involved in some of the new programs, but you cannot generate enthusiastic support—there's no rhythm to march to.

The greatest challenge in transforming any organization is not with your direct reports; the greatest challenge lies in generating new perceptions and behaviors from your middle managers. How could we take the same type of learning that was bringing change to the senior management team, add a level of excitement, and move the resulting knowledge and energy throughout the entire organization?

Leading Change is about exciting everyone to march to the same rhythms in the same parade.

CREATING THE PROGRAM

To get the attention, support, and enthusiasm of middle managers, we needed to create a program that would be an *experience* for people, something they would remember and that would influence their thinking for a long time.

This could not be just another training session that people would take and forget. It was much more fundamental to the successful transformation of the company.

The underlying purpose was to convince National's managers that change was essential, that change was coming, and that they were expected to be a contributing part of the process. I appreciated that they would need many new skills to follow through and succeed. My hope was that we could generate some fire and enthusiasm—let them see the parade, show them the steps, and let them share the pleasure of marching in it.

Not a typical or readily achievable goal in a large organization, but one that's absolutely essential to carry out the transformation from business-as-usual.

As soon as the urgent initial problems had been addressed, I organized the effort to create what would become the Leading Change program. The hammering, crafting, and implementation on any new effort is done by a team, but I always appoint one or two specific individuals to be in overall charge of the effort, directly responsible to me. For Leading Change, it was natural to select Kevin Wheeler, then manager of Human Resource Development, and David Kirjassoff, director of Business Systems Consulting. The two of them, aided by Emory Business School professor and National consultant Bob Miles, took the Leading Change concepts I had developed at earlier companies, and created a program tailored to the needs of National.

The name came out of planning session for the program, and even though I was leading that meeting I can no longer recall whether one individual participant deserves the credit. Regardless, National's innovation of the term "Leading Change" has since been widely copied, as testified by a recent book that takes the term for its title, and by an article in the *Harvard Business Review* (by John P. Kotter, March-April, 1995) which provides a list of steps for transformation that closely parallel some of the key techniques set out in our program.

The man who oversaw the creation of Leading Change was Richard Feller, whose title at the time was director of organizational excellence—the same title he held with me at Rockwell (and, incidentally, one of the very few people who followed me to National from Rockwell). A soft-spoken man, Richard has been powerfully effective in designing and running programs that produce real, lasting change.

Once the sessions began, Miles usually flew in from Atlanta to teach a case study on the transformation of the Rockwell division, and other case studies on General Electric and Lincoln Electric.[2]

THE FIVE ELEMENTS OF LEADING CHANGE

In order to gain the support and enthusiasm of middle managers, and equip them with the skills they would need, I challenged the Leading Change design team to create a program that would incorporate these five essential elements:

#1—Leadership

The word "Leading" in the title Leading Change is meant to convey guiding, directing; it's also meant to signify those charismatic qualities of a leader.

During my introductory period at National I had noted a lack of leadership among the company's middle managers, in a sense that goes beyond what the term usually encompasses. The idea here was to bring people to a higher level of contribution and productivity.

In business, the measure we use—the measure which every business leader and manager is ultimately judged on—comes down to dollars and cents. There's a linkage between the work that the ordinary department manager does every day, and the wealth-creation of the company. Managers who aren't fluent in the language of the financial statement cannot help but flounder in one of their major roles. Because they're unable to adequately understand how well their operation is performing, they cannot determine how to improve performance, and they're unable to provide guidance to their people in this area.

So the Leading Change program would need to deal extensively with financial issues such as how to read an income statement and a balance sheet. But not just the accounting aspects; rather, what we were after was the deeper meaning—how to understand what the financials can tell you about where you're succeeding, where you're not, and what you need to do about it.

In its final form, Leading Change students spend nearly half their time on issues of the business proposition.

#2—The Soft Skills

Most managers lack skill in the "soft" aspects of managing people. In a change process, a manager who has been helped to develop these soft skills will be able to understand the human factors when transformation creates upsetting reactions. I deal with the reality that most people react strongly when they find the things they've done for the last five or ten years being turned upside down. Resisting change is human nature; people do it to protect themselves.

By providing the National managers with soft skills, we would equip them to cope with the natural resistance and push-back that can bog down a transformation process or stymie it completely. (See Chapter 12.)

#3—Promoting the Vision

A major challenge to the design team was to make Leading Change an effective mechanism for getting people to understand our definition of success, and the importance of defining success. The program was to be the vehicle for getting people throughout the company to become contributors to the corporate Vision—helping to shape National's new definition of success. By contributing to the Vision statement, they would become intellectually and emotionally involved.

All of the company needs to march to the same drummer; the organization's leader is not that drummer—the Vision is.

#4—Building Cross-Company Connections

The Leading Change program opened the door to an opportunity for creating linkages across the corporation. These linkages are invaluable to any organization, but for a widespread company like National, they are difficult and costly to nurture.

We could have trained some Leading Change instructors, and sent them around to National facilities worldwide; instead, we brought people from the facilities into Northern California.

And we made a point of forming class groups of people who did not typically work together.

As a result, a production manager from National's plant in South Portland, Maine, seeking an answer to a question, might now call a test engineering manager in Cebu, the Philippines. Through their Leading Change class, these two each know people in completely different parts of the company whom they can turn to for support. We have extended the meaning and power of diversity.

Though the benefits are not immediately apparent, these linkages are critical to creating the kind of company we're aiming for—a company in which cross-discipline and cross-department efforts are routine, a company in which people can be innovative in finding ways to get a job done, instead of being bound by the strictures that fill a shelf of procedures manuals.

#5—Tapping the Right Brain

By their very nature, most production companies—especially here in Silicon Valley—are technical and engineering-oriented. In today's terminology, they are left-brain institutions. These are essential skills; a company like National couldn't survive without them. But there is a whole range of challenges in the workplace that can't be quantified, that can only be dealt with using right-brain, intuitive kinds of thinking—the hunch, the gut feel, something that tells you what you ought to do when you aren't able to gather clear reasons or justification.

We're trying to help people learn how to get in touch with their right-brain abilities and make this a more powerful dimension of their leadership skills.

In the Leading Change classes, this was never called "right-brain thinking," but the program was planned so it would specifically expose people to solving problems and thinking things through with right-brain skills.

This one aspect of Leading Change was so different from anything the company had ever done before that it proved memorable just by virtue of the contrast.

SUCCESS AT NATIONAL

Many managers who took advantage of the Leading Change program have found the opportunity to tell me how much this has meant to them. They often describe it as one of the most memorable experiences of their entire career.

I continue to be delighted by stories that National managers and executives share with me. For example, John Clark, the vice-president who is National's general counsel and corporate secretary, found that the most surprising aspect of the change process was the way the company's lawyers responded. John noted in a memo—

> The legal profession is highly resistant to change. From my discussions with peers at other companies, National is certainly not unique in seeing strong "push-back" from company attorneys on HR programs, quality initiatives, and similar efforts that have swept through corporate America in the last twenty-five years.
>
> The change process was initially no different than other similar programs tried at National in earlier years, in which lawyers professed to see no value. They consider these "touchy-feely," which is how they also categorized Leading Change when they first heard about it.
>
> After it was made clear that Leading Change was mandatory, all the lawyers went through the five-day program—and their attitudes afterward reflected a remarkable turnaround. The program was a "jump-start" in demonstrating to the attorneys that such an environment could contribute to their personal and professional success.
>
> A specific example of the change in approach is the continually increasing role of lawyers as active members of participative teams addressing projects or problems at National. They have committed to change, and are much more communicative, trusting, and team-oriented—and as a consequence more effective counselors to their clients and the company.

> The power and effectiveness of the change process was most graphically and surprisingly demonstrated to me through the effect of the Legal Department—in any company the most traditional and resistant to change.

For me it was amusing that another executive in an entirely different part of the company had a parallel experience. Dick Sanquini, senior vice president of business development, found himself saddled with the task of gaining support for the transformation from the company's intellectual property attorneys. In a session where he tried explaining to them a particular evaluation process that people around the company were using to analyze and improve their procedures, "sparks flew." But with the winds of transformation at his back, Dick put the patent attorneys together with the appropriate engineers and business people, and got them working jointly to improve the patent process. Each of the three groups assumed the others didn't even have the right vocabulary to hold an intelligent conversation with them. But given the task and the challenge, they all discovered—pretty much for the first time—that they could communicate with one another, and solve problems together. A revelation!

Three years later the intellectual property group had tripled the number of patent applications they were filing annually, ranking National fifth among all companies in California. Dick Sanquini could hardly believe the success. "Truly astounding," he called it. The patent attorneys admitted they could never have achieved this using their previous methods.

AN EVER-EXPANDING PROGRAM

We originally designed Leading Change for the direct reports of the forty people who had been attending the executive off-sites. These early sessions brought together groups of twenty-five to thirty people for five and one-half days starting on a Sunday evening, at some attractive and relaxing off-campus location where, we hoped, the participants would interact and share experiences and insights. Here's what the program covered; (a complete schedule of the week appears in Appendix B).

Sunday evening—National's new Vision, presented by a corporate vice president or other high-ranking executive.

Monday—"Leading Corporate Transformation": Includes case studies of transformation at Gil's Rockwell division and at General Electric, and a "visioning" exercise.

Tuesday—"Achieving Financial Success": Corporate treasurer or other financial executive on Financial Success at National; another case study; and issues and opportunities that govern the formation of National's strategy. (Afternoon session ends with a physical team-building activity.)

Wednesday—"Steps to Organizational Excellence": Presentations by outside consultant on Tools for Managing Personal Dynamics, and on Organizational Change Tools. Divide into teams to explore the Vision in depth, then present reports to the group. Repeat, on subject of implementing the Vision.

Thursday—"Exploring the Vision": Personal role in Leading Change and how to begin. Prepare plans on how each individual will implement the Vision in his or her own organization; rehearse, get feedback, and polish the plans for live presentation to Gil Amelio.

Friday—"Committing to Change": Two hours of live presentations to Gil, followed by remarks by Gil and Q&A. Afternoon: Communicating the Vision; evaluation and close.

The expectation was that we would put about 250 people through this training. But those who attended all seemed to leave with the plea, "My direct reports really need this." We found a way to include those people, too.

By late 1994, nearly 6,000 National managers and professionals—over one-quarter of the entire work force—had taken the course, many of them in a shortened, three-day version. Leading Change has been successful beyond anything we had a right to expect, and eventually every manager and supervisor in the company, and many non-management employees, will participate.

Adding to the original concept, we later developed "Pursuit of Excellence," a complementary, one-day program for the grass-roots level—assembly-line workers, first-level engineers, and so on, where they have a chance to hear about and make themselves heard on topics such as empowerment and how they could use it. Every National employee is receiving some kind of exposure designed to engage them in the excitement of the transformation.

The initial Leading Change efforts were so successful in launching the transformation that once we had achieved our initial set of goals, making the company once again viable, we launched a follow-up program designed to gain support for what we call "Phase Two" of the transformation, as described in the "Looking Ahead" section at the end of this book.

CREATING YOUR OWN LEADING CHANGE

It's a truism as well as a cliche: no important corporate program is likely to succeed without clear evidence that top management wholeheartedly supports the effort. In corporate lingo, the organization's leadership must "walk the talk."

A program like Leading Change won't make headway with limited support; even enthusiastic statements issuing from the executive suite or manager's office are not enough. Active participation by top management is essential if the program is to be effective and believable.

In our program at National, a Management Committee member always appeared at the Sunday evening opening session to address the group and explain the goals. The financial portion was taught by Sr. VP/CFO Donald Macleod, or when Donny wasn't available, by Controller Robert Mahoney or Treasurer David Dahmen. (Macleod had volunteered for the initial session, and the experience proved to be a rousing success—the participants got an eye-opening view of the company's financial situation, and Donny in turn was inspired by their enthusiasm. It was such a fulfilling experience for him, and clearly so valuable to the students, that he continued appearing in person whenever his schedule allowed.)

Most of the management team seemed taken by surprise when I mentioned that I'd be participating in the sessions myself. The first groups to go through Leading Change were people at or just below the vice-president/general manager level; I met with each of these groups at the end of their course—some twenty sessions in all.

The first part would be a presentation the group had jointly worked on, designed to bring to my attention any issues they felt were being overlooked. After that, I would speak informally for an hour and a half, addressing the issues they had raised, reviewing the company's situation, sharing my own views of what needed to be done, and challenging them with the crucial importance of their commitment.

Each group would also have prepared some questions, and they were encouraged to be tough on me. But as you would expect, the questions provided valuable lessons in the problems, frustrations, concerns, and hopes of our employees, and at the same time gave me a valuable cross-check on how the transformation was taking hold in the vital middle levels of the company.

It's easy for even the best-intentioned leader to be distracted by other pressing, last-minute priorities; there's a strong temptation to send a substitute to meet with groups like these. But only fire, flood, or pestilence should be allowed to keep you away.

The name "Leading Change" isn't sacrosanct, but there is no question that some program of this kind is essential if your transformation effort is to succeed.

You cannot sit back and expect the reasons and programs for change to percolate through the organization unaided. The understanding, enthusiasm, and support generated by an effort like Leading Change is a cornerstone of transformation.

WRAP UP

Most corporate change efforts are made up of valuable but distinct programs that are unrelated to one another.

To create the kind of fertile environment where change can flourish, you need to remove the speed bumps—the bureaucracy—and increase the steepness—promoting the positive stress, the feeling of "Can I keep up?"

The most productive departments in any company, and where employees feel most fulfilled, are usually those where the managers have nurtured this creative tension.

In the next chapter we continue this discussion of change with details of the program created at National to foster the rapid change needed in a transformation situation.

One surprise has been that the Leading Change program proved to be cross-cultural: it's working effectively in every country where we have operations.

Communication is the intellectual part of a program like Leading Change; the emotional part is pride. When people begin to experience a sense of pride, you not only get better work, you get better workers—people with brighter attitudes, who are sick less often, and who make more of a constructive contribution to the business. Pride gives people a sense of purpose and a feeling of being worthy. The effort validates individuals, and your company benefits from their improved quality of work. As a result, the transformation efforts extend far beyond your campus and factories.

The principal goal in designing a program like Leading Change is to ensure that no one will forget its central messages—that business as usual is no longer acceptable, and that every manager and supervisor has a role in bringing about change. Your design of the program must make every person feel valuable; they must each know precisely how they contribute to the transformation.

END NOTES

1. Drucker, Peter, *Harvard Business Review*, Sept-Oct 1994. Drucker postulates that a theory of the business has three parts: Assumptions about the environment of the organization, the society and its structure, the market, the customer, and technology; assumptions about the specific mission; and assumptions about the core competencies needed to accomplish the organization's mission. He insists that your theory of the business be tested constantly.

2. Case Studies—Rockwell (cited earlier): *Rockwell International Semiconductor Products Division*, Emory Business School case study OM88-101, prepared by harbridge House, Inc., 1988; available from the Case and Video series, Emory Business School, Atlanta, GA 30322.

General Electric: *General Electric*, 1984, Harvard Business School case study 9-385-315; available from HBS Case Services, Harvard Business School, Boston, MA 02163.

Lincoln Electric: *The Lincoln Electric Company*, Harvard Business School case study 376-028; available as above.

8

"Empowerment is Hell"

"Empowerment is something to be seized, not granted."

When Gil was a young manager running his first department, one of the engineers came to him and said, "You know, I can't even sign for a pencil and pad of paper in this company. When I need something, no matter how insignificant, I either buy it out of my own pocket or I've got to fill out some paperwork and get my supervisor's approval."

Gil found out it was absolutely true—those highly skilled engineers had no authority whatsoever to buy anything on behalf of the company. This added up to a situation in which there was no owner for the spending, and Gil could see that some of the engineers were getting ready to use what might be called the John Barrymore hat method. Barrymore, actually talking about an aspect of relations between the sexes in the days before the women's movement would have raked him over the coals for the remark, said a man having a problem with a woman should resolve it with his hat—"Take it and run." Politically incorrect by today's standards, it was, for the time, typical Barrymore.

Without buying authority, without a simple, straight-forward company expression of respect, Gil was about to watch several of the department's best engineers take their hats and run.

He introduced a new rule: every professional in the department could spend up to two-hundred dollars of company money on his own authority. They would have to document what they bought, and for what purpose, but they didn't need any approval.

Gil recalls what happened next—

> *Of course the company comptrollers immediately went into cardiac arrest, fueled by visions of unbridled spending. I told them, "if there's a problem, we'll change the procedure. Let's monitor the situation and see what happens."*

And something did happen. Spending went down.

Gil says, "This was a great surprise to everybody—even me."

Today he chuckles over the incident. But the experience taught him a valuable early lesson about empowerment.

> *The reason for this totally unexpected result was that people hadn't felt any responsibility, because giving them no authority sent the message that their opinions didn't matter.*
>
> *When they'd bring in paperwork for a twenty-five dollar item, the supervisor wouldn't want to waste any effort on it. He'd think, "this is insignificant, it's not worth my time," and he'd just sign automatically. A lot of times the engineers would end up putting in a request for things they didn't really need.*
>
> *But paperwork isn't a substitute for judgment. The new rule said in effect, "the buck stops at your desk for the money spent. Your spending has to make sense and it's going to be reviewed, but it's your responsibility."*
>
> *Suddenly people felt a whole different attitude about what they owned and what they were responsible for. And that's why spending went down.*

★★★★★

At National, as the employees began to get comfortable in the driver's seat of empowerment, they became a force for change. When Gil reached the question-and-

answer portion at one of his Leading Change appearances, he found the group had taken the invitation to "ask challenging questions" quite literally.

One of their questions was, "The top forty (executives and managers)—are they the ones who can lead us to the Vision? Those who are not walking the talk, how long before they are reassigned?"

How long before they are reassigned! The employees saw some of the executive team as dragging their feet, and were nudging Gil to get them out of the way.

But learning the lessons of empowerment doesn't always come easy; each person progresses at his or her own rate. If the managers perceived that some of their bosses weren't moving very fast, they would need to take a bit of Gil's own medicine: don't expect everything to change overnight, and when confronting two necessary steps that seem mutually contradictory, keep a clear head about discerning which has the higher priority.

The executives would stay for the outset... and Gil would continue applying pressure to make the messages of change, transformation, and empowerment start penetrating more rapidly.

THE NATURE OF EMPOWERMENT

Empowerment: The process of how you go about ensuring that Will and Hans, Maria and Yusef each know what their range of authority is, and know they have the freedom to take actions and make decisions within that range of authority. I think of empowerment as a way of amplifying myself: in the terms I described earlier, the question becomes—How could I create an organization where every manager facing a challenge would make the same decision I would make?

Everyone knows about empowerment, everyone acknowledges its importance. But I find that many people are either not truly convinced about its importance, or are convinced but unable to figure out how to make it happen. Or, still worse, *think* they have empowered their people when they're really continuing with the old management-as-usual.

This chapter is about how we practice "empowerment" at National Semiconductor, our steps for making it happen, and how these steps can be applied in other organizations.

EMPOWERMENT IN ACTION

When I arrived at National, Tom Odell was running the Analog Division, headquartered on our main campus in Santa Clara, and Kirk Pond was running the Digital Logic Division, with headquarters in South Portland, Maine.

Despite the fact that these were very large businesses—in excess of $300 million each—neither of these executives ran their own factories. They were dependent on a colleague, Jim Owens, who was responsible for all the plants in National. While this was fortunate for them in one sense, because Jim is a highly competent manager, it still meant that the division managers had to coordinate with someone else on matters concerning their own plants. Worse, because of the size and complexity of the job, Jim had twenty-two direct reports and very little time to be fully responsive to the division managers. If there ever was a job that was truly impossible, this was it. Looking for a better organization, I observed that we had basically two kinds of businesses: those that made "horizontal," or multi-market, products (items that many different customers purchased and built into their products) and "vertical" products (items made for just one application). These two require very different ways of going to market as well as different areas of emphasis in manufacturing.

It therefore made sense to divide the business units into two groups, each with several divisions. But who was I going to designate as the group president on the "horizontal" side, where Tom and Kirk were more or less equally qualified? And how would I ask one of them to report to the other?

When I'm considering an organizational change, I make it a point to talk openly about the fact. This signals people to get their ideas in to me and makes unnecessary the whispered conversations in the hall and the flood of e-mail messages. By making the idea of a reorganization the world's worst kept secret, nobody is surprised or caught off guard when it finally comes to pass.

Each of the several arrangements I considered for putting the two operations together seemed flawed. But in my struggle with the problem, I had been forgetting my own principle of empowerment.

The answer was to put the problem in their hands. When I did, after much work and effort they came back with an unexpected solution: they would integrate their two divisions, and in the process incorporate a third, and they would run this newly formed "Standard Products Group" as *co-presidents*. This was, to say the least, an unusual proposal, and one that all of the other executives were convinced wouldn't work. I had a few misgivings of my own. Nonetheless, it was their solution, and I had no strong reasons to oppose it; following the principle of empowerment, I told them to go ahead.

It turned out that the two men had very complimentary abilities and styles. Tom Odell proved to be strong on product and long range planning, Kirk Pond a "full steam ahead, get the product out this week" type of manager. The arrangement worked extremely well, lasting until I promoted Kirk to executive vice president and chief operating officer, three years later.

I've been asked about my handling of this situation, what a manager should do if the person or the people come back and say they can't find a solution. In fact, that did happen here: Kirk and Tom wrestled with the question for a while, and then announced that they were unable to find a satisfactory arrangement. I told them, "If I have to, I'll make a decision. But you'll be much happier with a solution of your own," and sent them off to try again. My rule of thumb is to give people three tries to do it themselves.

A solution handed down from the boss wouldn't have worked as effectively, because the people involved would have had no stake in devising it.

One of the most difficult aspects about pursuing empowerment is to convince people you really want them to work out solutions on their own. This one proved to be a good illustration for everyone else, as well.

MANAGING LESS

"Managing" in the traditional sense—controlling, monitoring, measuring, and so forth—are things that still have to be done, but they need to become less of a consuming focus for today's manager. What used to be fundamental to the definition of management, has become just one part of the job. In the era of employee empowerment, managers need to manage less, and lead more.

I believe every employee should be able to get things done on a self-initiated basis about 90 to 95 percent of the time. The ultimate objectives of the company can best be reached when the working atmosphere achieves a condition where everybody functions largely in a self-directed state.

To answer new and future demands, the manager must now become an educator, a counselor, but most of all, a leader. Managers do not have to be an expert in everything their employees do—and employees no longer expect it.

Instead, workers have a right to expect a professional manager with leadership, and with the judgment, strength, confidence and self-perception to grant empowerment.

(Another result is that we can "flatten" the organization because each manager can handle more direct reports when their people are empowered and self-directed.)

GUIDELINES VS. BUREAUCRACY: "DEBUREAUCRATIZING"

Suppose you're going to do some work in a city you've never been to before—but this particular city has been laid out by a very logical pioneer.

Streets run in one direction, avenues in the other, all in neat, parallel lines. The numbers start at the river, and there's a street sign on every corner. Wherever you want to go, you can figure out the way very quickly. In other words, the exact opposite of downtown Tokyo.

We can establish the same kinds of clear directional guidelines in our companies.

But there's a paradox here. You cannot have empowerment without guidelines, and guidelines easily turn into bureaucracy.

Bureaucracy is having a policeman on every corner verifying whether you should be on this street or not. You know exactly where you're going, but at every intersection you've got to stop, tell the policeman your destination, what you intend to do when you get there, and why that's a beneficial thing to be doing.

Some people stick to a rigid, rule-bound organizational style because the only alternative they can see is letting everybody do what they want, which clearly leads to chaos. Yet the opposite of bureaucracy need not be total freedom and chaos; it can be empowerment that comes attached to a clear set of guidelines.

Bureaucracy is a hindrance. Guidelines instead give you the information necessary to navigate efficiently throughout your organization.

You can have the freedom to make decisions, but you have to make them a certain way. You can drive your car in any manner you want, so long as you drive on the streets and not on the sidewalks or lawns.

Today's effective, empowering managers continually challenge themselves about the rules, procedures, and requirements placed on their workers. To discern the difference, put yourself in the place of a worker, and consider your own rules and procedures. Do they channel you in a direction appropriate to the Vision of the company and the goals of your work unit? Are they required by safety, law, or the ethics of business? Do they lead to decisions that are good for the business and the bottom line?

Or do they, on the other hand, discourage self-directed work and the search for better solutions? Do they quash innovation? Do they subtly announce that only "how we've always done it" or new ideas handed down from above will be acceptable?

Bureaucratic rules impede; an excess of freedom brings chaos; but guidelines *empower*.

THE SIX STEPS TO EMPOWERMENT

What are the things you need to do in order to become an empowered leader or manager?

The following are guidelines I've developed over a number of years, and which I have promoted heavily at National.

Understand the definition of success

> This brings us back to the company Vision, and the interpretations of that Vision done at each level throughout the organiza-

tion. Before you can accept the challenge of empowerment, you first need to be certain your efforts will be appropriate to the company's definition of success—the Vision statement.

Each manager must define the specific area they want to stake out—where, consistent with the Vision, they and their team will focus, and what, specifically, they will accomplish.

At the loftiest levels, this means asking yourself, and asking your employees, "What do you want to be famous for? How will your efforts change the world?"

Understand your status

To become empowered, you must first have a clear understanding of your status, both personally and within the organization—what are your strengths, what are your problems. From this base, you can build the confidence in your own ability to get things done... and then go on to make a difference. This is a topic that many managers gloss over when trying to empower their people, but it's a critical item that demands careful attention.

Develop metrics

Chapter 16 is devoted to the subject of metrics; here we apply the concept in a more individualized sense to empowerment. All managers need to establish metrics and benchmarks as a guide for themselves and their workgroups. Metrics enable you to measure progress toward meeting the goals.

Benchmarks, where appropriate, establish a base for measuring how you're doing against the *competition*. In some circumstances, it's useful to state the goal in *relative* terms; for example, I would like to see National's quality twice as good as next best competitor; (twice, because if only 10 percent, people would argue whether the figures were correct, whether we had indeed succeeded). By setting the benchmark in these terms, our goal is raised every time the next best competitor improves his quality.

Develop an overall roadmap

This calls for a clear expression of what you are trying to achieve, and how you intend to get there. The roadmap sets the path; it provides the intention, not necessarily the specifics.

Get buy-in from your team, and from management

Empowerment calls on people to recognize problems and seek solutions, and to put their own best ideas into action. This process doesn't get carried out in a cave; you need to solicit the ideas and advice of your colleagues, your subordinates, and your manager. At the same time, you should also be striving to make them part of the process.

Be accountable

Finally, you must be accountable. When you seek empowerment, you must be prepared to accept responsibility for the results.

I have told everyone in National Semiconductor, "Go at it—you can empower yourselves. The only ground rule is, follow the six step process. And if you do, you will be successful."

Many National people have responded with an attitude that essentially says, "This is great"; they've shifted into an empowered mode and have been every effective. Others went tiptoeing around, thinking that I didn't really mean it, or were not prepared to make the commitment and the change.

The people of National are making progress; we still have a long way to go.

APPLYING THE SIX STEPS

How do the Six Steps actually get translated into everyday reality? Suppose you're the manager of a wafer fab (semiconductor industry shorthand for a plant that produces semiconductor chips). The cycle time in your fab—the time it takes from the start of processing until chips go out the door—is eighty days.

Understand the definition of success: You know that your CEO talks about the need to reduce cycle times. An improvement in this area would therefore support an established definition of success of the company.

Understand your status: As plant manager, you have the authority to make changes without review, as long as they are within the budget and other guidelines you work under. Be aware of your own past weaknesses in carrying out major changes of this kind; then review the strengths that give you the confidence to succeed. For example, you might decide you need a more thorough staff analysis of alternatives than you have required in the past; or you might decide there are aspects you and your staff are not qualified to evaluate, for which you need to bring in a consultant.

Develop metrics and benchmarks: You know your current cycle time, and you have procedures in place for measuring. You will establish a benchmark by finding out what the cycle times of the competition are, and averages for the industry (see Chapter 16).

Develop an overall roadmap: As part of your planning, you will prepare a list of steps that might contribute to a reduction in cycle time; the list might include new machinery, exploring new technology, improved worker training, better computerized controls, designing a more effective workflow; even construction of a new plant may be worth exploring.

Get buy-in: You'll go around the plant asking workers, "What slows you down?" and "Do you have any ideas for improving cycle time?" You'll ask your supervisors, "What are the barriers that have kept cycle time from coming down?" Perhaps you've heard that the plant manager at another of your company's facilities has been doing a great job with this; even though your plant does wafer manufacturing and his does assembly, maybe some of his ideas will carry over, and you'll call him to discuss the problem.

Be accountable: Part way into the process, after looking over the initial information, you set a goal of reducing cycle time to sixty days. You launch the program that will start moving the plant toward the new goal. You are now spending money (probably) and effort (certainly) on these changes. The company will expect to see some benefit.

THE FLIP SIDE: RESPONSIBILITY

Empowerment comes with a string attached, and the string is labeled "responsibility." In the story related at the beginning of this chapter, about the engineers granted authority to sign for their own purchases, spending went down because the engineers understood that together with the authority came responsibility.

That's an essential message in empowerment: together with empowerment comes responsibility. When you empower employees to act, they must be prepared to take on responsibility for their actions. They need to understand clearly both sides of that coin. Empowerment must be exercised within guidelines that are specific, clear, and well understood.

You want to give your professional people a fair amount of latitude, removing the traditional straight-jackets, but this needs to be accompanied by great clarity about what they're responsible for.

SOMETHING TO BE SEIZED, NOT GRANTED

How do you transfer a sense of empowerment throughout an organization—to fifty people, or 1,000, or 50,000?

We try to get across the idea that empowerment is something to be *seized*, not granted. You empower yourself. The message is, "Go do something, go make something happen."

This is not a gift you give people; it is a power they take. People seize the opportunity when you create an environment, a culture, and the kind of organization that makes them want to be empowered—when they see empowerment as something essential to their success. It's the opposite of command and control.

What you're looking for is a kind of "organized chaos." A manager doesn't need to know every step his people carry out, he just needs to know they're following the system (the "process"). If the employees are empowered, and if everyone understands the principle, then it's okay for a worker to come up with a better solution—so long as he or she follows the rules. He has a license to make something happen and he doesn't need to worry about stepping on the manager's toes. And maybe he really does have a solution for something that's been a source of customer dissatisfaction. If your company's people have a clear and valuable end objective, and they follow the guidelines, they're apt to find ways of doing things the managers never dreamed of.

"EMPOWERMENT IS HELL"

Why is it that everybody talks about empowerment, but so few succeed? The answer lies in the phrase, "Empowerment is Hell." For those who understand the advantages but have no process for getting their people to embrace the idea, empowerment is a tantalizing but ultimately frustrating concept.

At the start of a transformation, your attitude is "I'm the teacher, you're the student." But as the transformation takes place, as people at all management levels become empowered, they must take on the dual role of student and teacher, simultaneously. You gather disciples—first preaching the gospel yourself, and then getting others to preach it.

Early in my career, I thought, "These people are smart, I do not have to tell them more than once." But smart people can find more sophisticated ways of missing the point. You must be even more repetitive and consistent.

Empowerment is a frightening concept, because most American companies have not taught people how to take self-initiating action.

But now you have a process that can make empowerment a reality.

WRAP UP

A major consequence of achieving empowerment is that organizational change can then be driven increasingly by the people (the "grass roots") and less by senior management. It's this aspect which makes it an enduring virtue: long after you have moved on, the organization has the ability to stay on top.

The "Six Steps to Empowerment" provide the framework for turning the concept of empowerment into an effective tool.

When a subordinate comes to you who has carried out all the steps—get out of his or her way. He may think you have empowered him; in reality, he's empowered himself. You know you're beginning to succeed the first time you get that welcome surprise when someone comes in to tell you about an initiative he or she has taken and, to your delight, it's exactly what you would have wanted them to do—they beat you to it!

Those who follow the system can be quite successful. And you can get people with vastly different backgrounds and ways of thinking about things, who are nonetheless consistently following the same set of guidelines.

The challenge is how you make people sensitive to the balancing act between freedom and responsibility. For a leader or manager, this tradeoff is the hardest part of empowerment to convey.

Remember above all that empowerment is something to be seized, not granted.

9

Encouraging Individual Performance

"To raise the level of individual performance, you aim to create an eternal restlessness in your management leaders. The best managers respond positively to change; leaders drive it."

In 1992, National Semiconductor launched a process of employee surveys, but of a very special kind: run and reviewed and summarized by the employees themselves. Gil Amelio strongly believes that an employee survey, to be of any believable value, must be employee-driven, and he insists that the team which analyzes the survey findings be made up entirely of grass-roots workers—"no managers welcome."

After the 1992 data had been collected, a diverse group of approximately twenty employees was recruited from operations throughout the company to analyze and interpret the results, and then present their recommendations to management. The members would meet with Gil and his senior executives to summarize the messages the employees were sending, and to offer specific recommendations.

Surprising as it may seem, Gil's established rule for this survey process includes a promise that the company will accept **every** *recommendation that doesn't violate the law, is humanly possible to do and is financially responsible. (Gil has used this technique in other circumstances and says that the results "never cease to amaze me.")*

Selected to join the assessment group at that very first National survey in the Amelio style was a woman we'll call Alice—an hourly, near-minimum-wage production worker from the wafer fab in Texas. Despite her limited formal education, her never having traveled far from her Texas birthplace, and her narrow view of the world and society, the assignment would require her to travel to California, interact with people of diverse backgrounds, and spend her days cooped up in meetings in corporate conference rooms. Suddenly she was thrust onto this whole new stage where the pace, content and action would be foreign to her experience.

She walked in shyly for that first day's session, unsure of what she might possibly contribute. By the end of the day, from the atmosphere of acceptance and mutual respect, a change was already becoming evident in the way she felt about herself. She spoke up with observations representing the viewpoint that she and her coworkers held, and could see that the group welcomed her input; her confidence grew.

Other participants noticed that her dress, posture, and voice seemed to match the growing confidence, which continued to improve over the two weeks.

Many of Gil's business ideas have a way of developing individuals; those who know him best feel that this is without doubt part of his every plan.

At the final session, Alice was one of the people who volunteered to stand in front of the group and present some findings and recommendations to her CEO and his top managers, including her own boss's boss. Everyone remembers that day as a watershed for Alice: it was clear to all that she had been a valuable contributor to the group. Back in Texas she would have some unbelievable stories to tell.

The company staged a cocktail party and going-away dinner for the survey team and the executives, and Alice, who had never before spoken with any company official above the level of her own supervisor, sought out Gil Amelio, to share with him how the experience had raised her to new heights. She had even surprised herself, she said, on how much she could contribute. Ideas were drawn from deep inside herself, in a way she had never been challenged to do before. She had tapped deep-rooted capabilities, competencies, knowledge, and experience.

Gil saw her again during a tour of the Arlington facility about a year later and was fascinated and rewarded to see that the changes had stuck.

Based on recommendations from that original survey group, National instituted programs that would make the company more responsive to employees' needs—among them an ombudsman program and changes in some of the physical work conditions. The company also agreed to provide better information on the principles behind how compensation is determined, to give workers a much better sense that there is a basis of logic and equity at the root of the system. Even at the corporate level, Gil recognized very real changes—

> *The consequence of this honest and fair system was people more devoted to the success of the company, putting out that extra effort, working that extra hour, finding that extra problem and solving it. The rate of change was dramatically and clearly increased.*
>
> *Almost every indicator showed improvement—scrap rate going down dramatically, yields going up dramatically, cycle times improving dramatically, the mere lack of mistakes—virtually every measure that matters to the company's productivity and health—because people now cared a little bit more than they had before, and it's very evident in the data.*
>
> *So, the on-going consequence to the company is more profitability.*

Clearly Gil doesn't measure success of changes only at the corporate level. He's aware that the Alices of a company grow significantly from these experiences.

Gil says, "An opportunity like this fundamentally changes the lives of every person involved. We've seen people going back to school, reshaping their basic attitudes, taking an active part in community help programs."

"Yes," he says, "there are rewards beyond the bottom line."

PUSHING DECISION-MAKING DOWN THE LINE

A *New Yorker* magazine[1] journalist, praising White House Chief of Staff Leon Panetta on his decisive management style, wrote that "... If you present decisions in an organized context and give him a clear description of the problems and the issues, with options that are tight, not loose, he'll ask questions and then make a decision."

Most people would see that as effective decision-making, and applaud it. I applaud it, too... but only faintly. Your goal as a leader should be to find ways of encouraging your people to make decisions themselves.

A favorite adage of space pioneer Werner von Braun was the old Chinese adage that he quoted this way: "Give a boy a fish and you feed him for a day; teach a boy to fish and you feed him for a lifetime." Making decisions works the same way. The goal is to teach people so they won't have to keep coming back to you for "fish" or for decisions.

This involves turning your executives, managers, and supervisors into *risk-takers*, within the business model of the enterprise.

At National, this aspect of the transition process was especially difficult because, as I've made clear, it required a significant change in the corporate culture. People had learned from my predecessor Charlie Sporck to pass decision-making up the line—the standard workable management style of an earlier era. I needed to break the mold, to create an environment where everyone would learn to take responsibility themselves and to pass decision-making down the line.

One example of achieving this goal was with a National Senior Vice-president I'll call Andy. Early in my tenure, Andy came to me to explain how the problems with quality of a particular product were damaging our working relationship with an important customer. He presented the facts in an organized and typically articulate way but then paused respectfully, waiting for my instructions. The short wait stretched into a long, uncomfortable silence. Andy finally broke the silence and asked, "Well, Gil, what do you want me to do?"

My response was to ask him for *his* solution. It turned out that he had been considering the possibilities. He had a fine solution, and while it wasn't exactly what I would have chosen to do, it was, indeed, a reasonable approach. I nodded and smiled, and said, "Sounds good—work up a plan and I'll give you the go-ahead."

Very soon after that, Andy came to me to share another problem and, as you might expect, the conversation followed the identical pattern. At the third incident a few weeks later, I noticed that the quiet space was a little shorter in duration and not at all uncomfortable. Andy had come prepared with a solution, and he had the outline of a plan with him ready for me to review. During the fourth encounter in problem solving, Andy stopped himself part way through, and said, " Okay, Gil, I know—you're waiting to hear my assessment and my solution." I nodded, we smiled; the transition had taken place.

I knew what my goal was for Andy, for our relationship, and for National; I had worked it through with him in a patient way that would build his confidence rather than tear him down with criticism.

Since then, this VP has never come to ask me for a solution. Along with each problem, he brings a recommended solution and a process ready to present for my okay. This development has influenced his other leadership skills, and he has learned to affect the behavior of people on his team in this same way.

I've encouraged all of National's top managers to think beyond problem identification into the area of solution planning. Most now move through those meetings without missing a beat, accomplishing more and feeling validated by contributing to the limits of their responsibility.

The exception is the occasional manager you come across who exhibits a self-confident solution style from the first. An example at National is Kirk Pond, who automatically operates in a decision-making mode; I believe he developed that solutions skill-set because he was based across the continent, in Maine, and either made those crucial decisions and designed a process that worked or nothing would happen. Not by coincidence, since June, 1994, Kirk has been National's COO.

MODUS OPERANDI

How do you pass these skills along to your direct reports?

High quality performance is best encouraged when you give people a precise set of boundary conditions; you tell them, "Here are the limits of what you can do...but I expect you to work and function to the full extent of those limits—not to stop short and play it safe." In other words, identifying problems and creating solution possibilities are part of the same function. Notice how many people are eager to point out the problems but unwilling to create solution possibilities. Some people are all too fast to give solution ideas but are unwilling to create a plan and then a process needed for follow through.

When people bring you a problem, be sure they bring a solution plan with it. This plan should present a well-thought-through *process*, and should include—

- a clear description of the problem that needs to be solved
- the plan to solve the problem
- how this supports the goals or priorities for the work group, division, or company
- what resources will be required
- what the steps in the process will be
- what the milestones will be—with *specific completion dates*
- dates for presenting progress reports; and,
- what the measure of success will be—how they and you will know if they have succeeded.

Your experience will let you identify and modify any flaws in the process and offer important suggestions, when needed, to reinforce the plans. Once you're sure that there's a logical process in place, you can set your people free to pursue their plans.

I believe that intelligent people *always* have smart ideas—that is not the problem. You must encourage them to add practical plans and powerful processes to these ideas. If you do succeed, you can be assured of many successes for your company. And your people will have grown into solutions-minded, process-oriented thinkers.

THE FULLER ORGANIZATION

What I call a "fuller organization" is one where the individual's value is skill-based, not position-based. A manager or worker is valuable to the company because of what they know, their dedication, their experience, their wisdom, and not because they happen to be in a particular job. They're playing first base today, but because they're a member of the team and have certain competencies, I may want them on third base tomorrow.

In today's reality, you must be equipping people to do more than just one task. Otherwise you are dooming them to a dark and uncertain future.

In Chapter 11, we talk about organizational structure for today's corporation. One essential element of this is to ensure that we have a fluid organization, which underscores the need to be skill-based, and makes skill more important than organizational position.

ADVISORS AND INFLUENCERS

And where do these increased skills come from? Part of the answer is, from the outside—from advisors and "influencers."

Most creativity happens when you submerge your people in a sea of many inputs from diverse areas. As a consequence, they're able to recognize relationships that others have not seen before, and boom—you have a patent, or a marketable idea, or a better work method. The ability for people to be innovative in that way goes up almost exponentially with the ability to keep them in touch with what's going on beyond their own circumscribed sphere. These outside activities provide perspective and a frame of reference.

So to create a high-performance, high-standard organization, you need to find ways of providing these outside influences. Some of the ways you can do this are—

- bringing in university faculty and other experts to speak about the frontiers of knowledge in their field
- allowing people to attend classes and stay current on their education
- encouraging people to attend conferences in their field, and to take part in the local chapter of industry associations
- providing opportunities for attending the more worthwhile lectures and seminars in management.

I went to an early one of Tom Peters' executive sessions that he calls "Skunk Camps," and while I didn't learn much of anything new (the sessions covered what was in his book, which I had already read), people I met during the two and a half days provided me with plenty of "takeaways" that I'm still using. One old acquaintance shined in a new light:

Tom Malone, of Milliken & Company, Inc., a classmate from Georgia Tech who in a sense I rediscovered at the session, turned out to be someone from whom I've learned a great deal over the years—such as the idea of "Sharing Rallies," which gets grass roots employees actively engaged in transformation by bringing teams together to share ideas that have brought success in their work area.

Another was Frank Perdue of Perdue's Chickens. I suppose that might be good for a chuckle: what could the CEO of a Fortune 500 high-tech company learn about management from a chicken man? The answer is, plenty! In particular, the idea that what we need to deliver to a customer is not just a product, but value, is a concept I learned from Frank. (I hope he decides to write his own book on management, sharing his insights with many others.)

Most of what I have learned about management, other than from direct experience, has been in situations like the Skunk Camp, which provided benefits I could not have anticipated.

In line with this kind of "extra-curricular" learning, I applaud when a worker in one area has a curiosity to learn about something going on in another area down the hall. I look for ways to encourage this "out of cage experience" that can lead to unanticipated results. At National we're encouraging this with a program that's specifically designed to nurture efforts which break the bounds of the organization chart and project-based rigidity. We call it the "100/100 Program."

THE 100/100 PROGRAM

As another thrust of encouraging individual performance, we beat the bushes for National engineers who have a product idea—which does not have to be restricted to the area bounded by the person's department or job description.

The idea for 100/100 was suggested by National consultant Mike Townsend, based on a program used at Fluke Corporation, and was developed by National manager Demetris Parakevopoulos, who visited several companies renowned for their programs encouraging innovation—including Hewlett-Packard, Corning, 3M, Raychem, Applied Materials, and

AMP. A brochure created to promote the program stokes interest by reminding employees of National's "long-standing tradition of empowering employees to create patentable technology solutions that keep us in the forefront of a competitive global market."

An employee with a hot product idea begins the process by filling out a form that asks for a description of the product, its uniqueness and special advantages, how it would be manufactured, who the potential customers would be and how many of them there are, and the anticipated revenues. Of course, few employees are in a position to supply all this information, and so we provide the help of an "advocate," who guides the applicant as needed, including putting him or her in touch with experts in marketing, manufacturing, and appropriate technology fields.

Once the application is completed, it undergoes a screening by a preliminary panel of managers and directors, who evaluate it on the basis of criteria that include—

- technical feasibility
- meeting a critical customer need
- being closely aligned with National's core competencies
- offering significant advantages over competitors
- providing a large revenue opportunity compared to the investment
- requiring a distribution channel in which National is already a player, and,
- providing a strategic fit with National's targeted market segments.

Ideas deemed worthy of further consideration are returned to the originator for completion of a more elaborate form, expanding on the previous data and leading to the development of a plan covering the technical and business aspects of the proposal.

The plan then goes before a Review Board headed by our senior vice president for corporate technology, Charlie Carinalli. Only about one proposal in ten has made it all the way through the process and been funded, but for the successful applicant, the reward is a budget of up to $100,000.

There's a catch, however: the successful applicant has 100 days to create a working prototype or model... but must do this on his or her own time. You might think that would demolish all interest. On the contrary—Jewel Savadelis, who is now the administrator for the 100/100 program, finds that the kinds of employees who push their ideas through this process are people who "have a fire in their belly, a real drive to see their idea come to life." They are, she says, "people who would pay for the opportunity." Five months of working nights, weekends, and whatever time they can coax from their bosses, is something they willingly give.

But if the idea turns into a product, the rewards can be significant. The innovator may have the opportunity of becoming product manager or to take some other major role in the new venture. And he or she may garner an entrepreneur's status, risking some income against the possibility of a take-home pay that could reach five times the usual salary.

As of this writing, National's program is too new to proclaim it successful, but it has generated a lot of enthusiasm. There are several likely-looking products in the pipeline, and we are enthusiastically promoting participation by more employees.

LEARNING TO LIVE WITH MAVERICKS

Most of what we deal with in these pages involves managing the "typical" employee. What about the non-conformists—the mavericks, who are often embraced with as much warmth as a carrier of the plague.

Management consultant Arynne Simon[2] takes exception to the usual view of the maverick; she says most people think of a maverick as someone who "works odd hours, often coming in late but staying half the night; who comes late to meetings; who refuses to follow instructions; and who can't live within the confines of rules." Dr. Simon observes that managers and co-workers alike too frequently see the manifestations but miss the significance.

Instead, she describes a maverick as "someone who latches onto a dream and goes at it with such passion that they have blinders on about everything else. Because of their intensity, other people become impatient with them." The true maverick will put his or her own career aside to see a project through—perhaps passing up a promotion or making some other decisions that seem self-defeating.

But mavericks, because of their intensity, are very likely to be successful in driving their projects through to completion, and so have the potential of being a highly valuable asset to the company. A leader's task, Arynne believes, is to get the mavericks to buy into the Vision of the organization; once they have adapted their own vision to include and support the Vision of the business, they become invaluable.

Generally, mavericks should not be squelched, or forced to comply or leave; they bring a wonderful spark to an organization or team.

How do you encourage them without letting them run out of control?

A response of "No" to a maverick could cause you to miss the nub of a valuable idea and inhibit some valuable thinking. Instead of taking that track, I've learned to lean back and listen calmly, asking a few thoughtful questions that might reveal the scope and depth of the idea. Then I ask for a plan—because, again, this encourages process thinking. I see my job as encouraging mavericks to turn their ideas into valid business propositions. Once they've done that and we've agreed on the process and the timeline and the measurables, I let them charge ahead with my blessings.

HANDLING PEOPLE WHO RESIST

A true maverick is usually willing to adapt to new leadership; it's often the people without spirit who seem to resist most belligerently. While mavericks can usually be redirected to channel their ideas into productive programs, every now and then you encounter someone who says "Yes" and doesn't do it, or openly challenges you.

When I took over the Semiconductor Products Division of Rockwell, I found that two of the executives resisted any new approach. They took a stand and firmly planted their feet. The odds seemed against moving them off the spot. My ideas were going in one ear and out the other, yet these executives were unable to recognize and acknowledge that behavior. This was feet-firmly-planted status quo, with zero prospect of change.

It became clear that I would have to be up front about what would happen if they continued not to listen. I dislike confronting grown executives in this way, but I sat down with each of them and said, in essence, "I came here to fix this business. With your help, I'll do it faster, but I'll do it with or without you. We don't have time to waste. The choice is yours. Let me know what you plan to do."

One of these executives, Charlie Kovac, had been with Rockwell for twenty-five years, and thought he was invulnerable. But he went home after my conversation with him, had a talk with his wife, Milly, and, came back to tell me that he understood and was prepared to change. "You've got my cooperation."

Charlie then proceeded to make the most rapid change I've ever seen take place in a powerfully experienced senior executive.

Describing that period of conflict, Charlie still remembers, "The more I followed Gil's road map, the more freedom I got." In fact, he turned out to be a crucial help in transforming the Rockwell division and became one of my most faithful disciples for change; people would go to him in order to understand my intention—"What does Gil mean by this? What does he want us to do here?" Charlie's grasp of my transformation principles became so much a part of him that he continues to serve as a consultant for me today, and remains a personal friend.[3]

When I sat down with the other executive—whom I'll call Jack to protect the identity of the guilty—his attitude was, "We've had a new president of this business every year for the last three years." The implication was obvious: "Why should I bother doing things your way—you'll be gone in a year."

Jack was vice president of operations, running a division that someone had nicknamed "Fort Ops" because they acted as if they were surrounded by barbed wire. They didn't cooperate with engineering, they didn't cooperate with marketing, or with anybody else. The attitude was "We're doing the best we can, it's someone else's problem."

I know that one aspect of my personality may be confusing because it cuts both ways. If I think someone is bright but not trying, I become impatient very quickly. But when I feel a person has the capability and is making a real effort, I show an enormous amount of patience. I'm sometimes faulted by my top managers because there is such a thing as too much patience. I recognize that I have a tendency to ride too long with people who show promise but aren't delivering—trying them in different jobs, and spending extra time on them until finally recognizing I should have moved the person to a position of less responsibility months earlier. That's the professor in me, wanting everyone to graduate. And I have to watch that tendency carefully to be sure I don't turn a virtue into a liability.

But in Jack's case, months went by without change, and his recalcitrance was a signal to others that it was possible to stonewall the Amelio changes and get away with it.

Corporate leaders come in as wide a range of personalities as any other group in the society; management styles vary from one extreme to the other. Despite the variation in styles, one fundamental I have always believed should be followed by all leaders is that you should not lose your temper. The leader should be someone with the discipline to control over-reactive behavior. A leader should be someone who can be counted on to solve problems and situations intellectually.

That doesn't mean you can never let any emotion show; I believe there are occasions when anyone is justified in getting heated—so long as we're not all heated at the same time.

Anger that leads to a total loss of control is in a different category. Good leaders keep themselves from showing anger; calm and self-assured behavior is an indication to everyone that the brain is appropriately at work.

But there's an exception to every rule. I've learned that sometimes when I want to send a message, I need to allow myself to "lose my cool"—in a precise and carefully calculated way. Anger is acceptable as long as it does not cross the line into over-reactive behavior.

Jack provided one of those rare occasions when I deemed this tactic justified. I decided to show my anger and disappointment, and I fired him. I did it publicly, using none of the usual excuses about "early retirement" or "more time with the children." Every other executive got the message: if you didn't want Gil Amelio's changes, you didn't have to accept them, but you wouldn't have a job for long.

If it's necessary to shoot one of the lead buffaloes in order to send a message to the rest of the herd, you'd better be prepared to do it before the whole herd runs off the edge of the cliff.

Again, this way of handling problem employees must be the exception. Since the Jack the Resistor days, I've evolved a personal procedure for this type of situation, based on my conclusion (which is clearly not original) that when trying to alter behavior, you should start with the carrot before wielding the stick, since most people respond best to *positive* reasons for changing behavior.

So, faced with a resistor, I put on pressure in an escalating series of steps. And what I begin with may surprise many people: I call for a "360-degree review"—an evaluation by me (or, if it's not someone who reports to me, then by the person's immediate superior), by peers, and by subordinates. The value of this is that it eliminates deniability—the person can no longer tell themselves, "The boss keeps griping, I have a problem with this, but it's really his problem, not mine." In depersonalizing the issue, it makes the problem much more believable.

Once the person has been faced with the results of the review, I then ask for a *written* program for correcting the problem. And I make it clear that I expect to see definite progress on the points outlined in the program. Most people will by this time have heard the message very clearly, and will be making a determined effort. Typically I'm willing to allow a month or two for distinct signs of improvement.

If the rate of improvement is still hovering around zero after this time, I then put the person on warning that they are facing the loss of their job. But in fact, it should be rare that you ever reach this stage.

THE NEW MEANING OF LOYALTY

One of the elements at root in all this is what we mean by loyalty, and I believe the definition of loyalty needs to be reconsidered for today's business climate.

Question: If two workers are each doing exactly the same quality of acceptable work, which is the more loyal:

Worker A, who has his/her office so filled with personal belongings and decorated with collectibles that it creates an aura of forever. This person talks about being dedicated to the company, and the company is obviously top priority over all other values. You know that Worker A would never entertain a job offer from another company, and probably even refuses phone calls from executive-search firms.

Or:

Worker B, who has set out a few family photos and meaningful books, looks at his or her position as one step along the path of a successful career, and makes no secret of having from time to time allowed an executive search firm to set up an interview with another company.

Worker B has created options but stays as a matter of choice; workers like A stay because of devotion and perhaps because they've closed off other opportunities. These are both loyal workers but I prefer the person who, despite being offered other options, chooses to stay. The variety of loyalty offered by frightened people does not impress me. Workers like B stay with your company because, after looking at options, they remain convinced that your company is where they want to be.

Loyalty is a conscious choice. Companies must work to win the respect and loyalty of their people. In America, I believe we should strive for a working environment of mutual respect and trust—where people choose to be loyal to companies whose leaders behave in ways that nurture loyalty. Company ethics should reflect a morality of leaders who behave respectfully and courageously towards their people—not just adhering to government regulations and restrictions but acting out of personal conviction.

Companies that invest in the development and training of their people are exhibiting respect and working to prove that "this company is worthy of your loyalty." With their enhanced skills, your employees can go down the street and work even more productively for another employer. You must be able to take that risk by working to win the much desired loyalty. We know that we invest and work mighty hard to protect our installed base of loyal customers; my advice is that you work just as hard to establish the same attitude towards employees. Your actions should announce, "There are no guarantees in either direction, but we want you to stay, we're going to work as a company to show that we respect you, and that we are, in turn, worthy of your respect. And we want employees to have that same attitude towards the company."

The 21st century is nearly upon us; we need to evolve and update many of our outmoded definitions and ideas.

John Sculley once caused waves throughout Apple Computer when he told the employees, "Don't make Apple your life." People across the company thought they were being warned of a mass redeployment, which did not happen. Sculley risked losing the loyalty of his work force but he set up a career center where trainers explained his theory of employee independence. He was way ahead of his time.

Redefining loyalty is tougher than just repeating the same old nuggets and having everyone agree with you. Sculley was right; the loyal workers aren't the ones who look on the company as a lifetime commitment. Instead, they're the ones who are with you because they have decided this is where they want to be; like restating the marriage vows, they keep their eyes open to the options, and continually choose, after weighing the options, to stay with you.

If you want loyalty—grow your people, develop them, help them to be strong, and they will likely choose to remain with you.

THE LEARNING LOOP: FEEDBACK IN LEARNING

A good deal has been written on the learning organization and so I was tempted to skip the topic here. But it's a concept too important to ignore, and especially important during a transformation effort.

First of all, how do people learn? If you stick your hand in the fire, it hurts; your nervous system gives you feedback, and you change behavior—you're not going to do that again.

What happens in an organization? Since an organization doesn't have a nervous system per se, if you want learning to take place, you've got to put feedback loops in place. Feedback loops are enabled by managers; if managers don't create them, they don't exist.

In National Semiconductor we make wafers in South Portland, Maine, that we ship to Penang, Malaysia, where they're assembled (our plant is, coincidentally, right next door to AMD and a few blocks from Intel). From the assembly plant, the finished chips are shipped to the customer.

Let's say a batch of these chips go to NEC in Japan—one of our main customers there. NEC gets the chips and telephones their National field sales engineer. "Take this garbage away," NEC says. "The reject rate is one percent and we can't tolerate that. And you've got our line shut down, so get us some good ones in a hurry."

Our sales engineer, being a dedicated worker, hurries over, picks up the chips, and drives them straight to our test center outside Tokyo. They rescreen the parts, sort the good from the bad, and the rep runs the good ones back to NEC. The customer is happy again.

But what happens next month? Very likely, a repetition of the same story. A few more times and we'll lose the customer.

How should this scenario actually run? After the rep has delivered the good chips to NEC, he sends the bad ones back to Penang, where they examine them and make some changes in the process.

The next month perhaps NEC calls up and says, "The defect rate is down to 0.1 percent—still not good enough, but you're making progress." Again the rep sends the bad parts back, and Penang make some more changes. Month by month, the quality improves.

In this process, there are *cycles of learning* taking place. The effectiveness of the learning is determined by how fast and how many times one goes through the loop.

The limiting factor here is cycle time. When an organization can navigate around that loop quickly, learning takes place rapidly. The goal is for your people to learn faster than your competition. Regardless of where you started and where the competition started, if you're learning faster, you will eventually pass them.

I say we had better learn how to learn faster than the competition.

Short cycle time brings other advantages, of course, but the chief one is to accelerate the rate of learning. If your managers succeed in accomplishing that, your company will ultimately win.

"NATIONAL SEMICONDUCTOR UNIVERSITY"— A VIRTUAL UNIVERSITY FOR EMPLOYEES

To enhance learning at National in a variety of forms, in 1994 I chartered a team of senior human resources people—Kevin Wheeler, Richard Feller, and Bob McLean—to create a high-level in-house educational facility. The result, which we call National Semiconductor University, held its first class eight months later.

What I had envisioned, and what the team has created, is something that goes well beyond the skills and professional training that are the usual terrain of most in-house training departments; these are of course also included in the NSU charter, but I was looking for something more, something that would help make our entire company a *learning* organization, developing employees who are well prepared for and skilled at driving change.

NSU was to be an organization that would do more than just provide information. I envisioned an organization that would be dedicated to encouraging the free exchange of ideas and best practices on a global scale. If the programs would enhance the "connectedness" of National people globally, and at the same time stimulate the creative juices, I would judge it successful.

The founding principle is this:

> *The best education is achieved through the process of discovery.*

Kevin Wheeler, now the director of NSU, says, "This is an exciting time for us, because we've been given a charter that allows us to try so many different kinds of approaches." He describes the pattern as being based on "the way Gil Amelio dealt with his direct reports from the first, in the way he educated them, gave them time to reflect, got them talking about issues that are not P&L oriented."

In some of their barrier-breaking approaches to education, the NSU people seek to go beyond the instructor-and-whiteboard session, getting beneath the surface to the underlying assumptions.

It's certainly nontraditional in a corporation to label as "education" an effort like this: Consider for a moment that a manufacturing process has been transferred to a new location, and it isn't producing at the projected rate. The NSU people look on this as a learning opportunity; they get a group together to probe what went wrong, to capture the lessons of the experience, and to make them widely known throughout the company. Problem solved; people developed.

Another approach brings together design people and manufacturing people, and throws them a topic like: "You guys are not designing stuff we can build." An open yet directed discussion ensues. The hope is that the designers start doing things differently while everyone learns to communicate, negotiate, and mediate.

NSU has also built close affiliations with several universities—including MIT, Georgia Tech (my alma mater), Santa Clara University, University of Southern California, Stanford, and Berkeley. Faculty members are brought in to give under-graduate and graduate courses that National employees can attend on a combination of company and personal time, and without ever leaving the building. If calling this operation a "university" seems like overstating reality, NSU students can actually get an AA, Bachelor or Master's degree from San Jose State University, the University of San Francisco, or one of our other partner institutions. Currently about 300 students are enrolled in 150 courses, in subject categories ranging from Business and Manufacturing, to Leadership Development, to Total Quality, to Technology Development. The tuition costs for these students are being picked up by the company.

In an unusual arrangement, these programs are also open to non-National employees—pursuing the goal of combining the academic with the real world, and at the same time making us a more responsible part of the local community.

The spirit behind the words "life-long learning" permeates these classrooms. Even people who leave the company part way through a course are permitted to finish, because we want them to have a warm feeling about moving on.

Professors and consultants are also brought in to lecture on their specialty fields, to keep National people abreast of the latest concepts in technology, administration, and management.

Motorola has been a pioneer with efforts of this kind, and General Electric has for some time been using a corporate function to drive change. Probably the closest to the National model, though, is Quaker Oats University, an operation that strives to eliminate the typical classroom, substituting on-site learning where practical problems can be faced and problems solved.

We spent months studying what everyone else was doing, designing and redesigning what we wanted to do before we finally launched this program, and the time spent on design is proving to have been well justified. I have probably gotten more E-mail from my employees on this than on anything else I have done at National. One message came from a lady who is a single parent; she wrote, "I thought with my personal situation I'd never, ever be able to finish my degree work and get that diploma, but now that I'm at National I know I can do that. Thank you very much." Not only is this program doing what we want to do in terms of offering skills, but it's proving to be a great way of retaining our critical people in a highly competitive industry.

These very new approaches to corporate education represent a trend that I believe is certain to spread. Any company can profit from a "university" that sets its goals higher than the traditional skills and professional training.

My standard is that a company should be spending about two percent of payroll for training. Increasing to this level should take place over a period of time, even as much as a couple of years, so you are wisely investing in programs that will work. As of 1995, at National we are currently spending 1.7 percent (even with the costs of NSU), and continuing to increase slowly.

Even a startup company can have one person on staff who studies other startups and arranges for an occasional visit from a professor or consultant. For a startup to spend 0.5 percent or 1 percent of its venture capital on this would prove well worth the cost and effort.

WRAP UP

Bring people along by encouraging them to make their own decisions. When appropriate, they can come to you not just with a problem for you to solve, but with the problem and their suggested solution. As your people learn how to present solutions to problems, they will upgrade the level of their input. In a short while they will put you in the position of saying, "Sounds good to me, go ahead."

END NOTES

1. Sidney Blumenthal, the *New Yorker*, July 11, 1994, p. 33.

2. Dr. Arynne Simon is a consultant on team building and communications skills to corporate leaders, especially in the computer industry, and is the author of the Simon Says newsletter. She is married to a coauthor of this book.

3. Commenting on how he changed as a result of working with Amelio at Rockwell, Kovac speaks of "how much more I liked myself after my eight years with Gil." Interview of November 4, 1994.

10

The Road to High Standards

"What you punish is not mistakes, but the failure to recognize and correct them."

According to business professor Bob Miles, Gil "believes passionately in the development of people," in raising their performance to levels they had probably not imagined. Miles says, "He creates an infrastructure for developing people."

Gil himself "can handle enormous complexity. He can keep lots of initiatives simultaneously—like a pilot monitoring all the instruments on the panel and running down a whole checklist before he acts."

This active intellect has a potential downside: Gil has a tendency to launch more ships than his admirals can control. But Miles notes that Gil, understanding the downside possibilities of his mental energies, has put mechanisms in place that help channel his zeal.

Miles is convinced that Gil will go down in business history as "the developmental CEO," and says, "He develops his people and keeps them longer than other CEOs. Perhaps it's having in place the process architecture for raising levels of performance that gives him the space for this development."

Gil's executive assistant Bonnie Murphy notes that Gil "never raises his voice, never blows up, never slips into profanity. But he admits to reaching a boiling point when he is convinced that someone with real ability won't try or when he sees someone make the same mistakes over and over again."

Bonnie says, and others agree, that "having Gil for a boss is not all sunshine and roses." She says—

> *Many of us have come to understand that he doesn't always tell you when he's pleased because this is what he expects. Although some may see this as being difficult to please, it's not that at all. Gil always sees something else that could have been done, which encourages you to stretch even more. He seeks a level of detail that's staggering. I would think I had gotten every possible detail, and he would come up with just one more.*

As a manager he became aware that most people are eager to raise their levels of knowledge and performance—that most people have a built-in desire to improve. And from individual improvement, the business proposition improves. But for Gil this is not just a business decision. To an unusual degree, Gil respects the intrinsic worth of people— their capacity to contribute and their desire to grow. His basic respect for the individual is his personal motivation to treat them with consideration. He is always shocked at the need to convince managers that development of their people is all important and personally rewarding as well.

Dave Kirjassoff, the director of organization performance at National, captures Gil's commitment in a nutshell: "He puts out a tremendous effort to make things work better, which sometimes surprises me. Four years ago, before Gil arrived, wires were hanging from the ceiling, people were being let go or finding other jobs. Today, people voluntarily work ungodly hours. And some of those who left seven or eight years ago are now beginning to come back."

That's transformation.

RAISING THE BAR

Recall my earlier example of the high school track coach; you want to start training your team in the high jump, where would you set the bar? If you set it too high, they'll get frustrated and turned off.

Instead you choose a lower setting. You want to teach the techniques, work on form, see what helps and what doesn't; in the process, the athletes will gain confidence. As they become able to handle the initial height, you gradually begin raising the bar. You inch it up a little at a time.

That example can be applied to business but it is too often overlooked. Typically we start people with tasks way above their capability. Perhaps we are trying to be respectful and are challenging others as we would like to be challenged. Generally, the better way is to set people up to succeed.

As an organizational leader, you would do well to emulate the track coach. Get your team together and set a short-term objective at a level that everyone is likely to be able to reach. Make sure this initial goal, although a stretch, isn't too ambitious—it must be doable.

Left to themselves, many people might safely stay in their comfort zone at the level of reliable achievement, and not because they can't or don't want to learn or be challenged. The truth is that most people are afraid of failure—or perhaps afraid of the embarrassment of failure. They will grow if you nudge them out of the comfort zone. Some management styles either leave people alone in safety to grow flabby and bored, or push them out into waters in which they might drown. I suggest a more practical and respectful way.

Like a coach analyzing the individual strengths of each member of the team, it's important for you to identify the comfort zone for each of your people—each one's easy strength, each individual resistance level when the going is getting tough—and you gradually raise the bar from that level. You start raising the challenges—all the while coaching, coaxing, and looking for ways to harmonize individual styles with the organizational style and purpose.

This is not a problem confined to young supervisors and managers who may tend to lack confidence because of inexperience. Like a flu bug that attacks people regardless of age, education, race, or gender, this one goes all the way up the line and can strike people of superb ability and lengthy experience. One example of many comes to mind: when I promoted Charlie Carinalli to become National's chief technology officer, I cautioned him that while he was a great, highly innovative product designer, the new position would require him to address the issues of *process* technology, as well. Intellectually he understood this challenge, yet he focused on the comfort-zone areas of product engineering, and found reasons to avoid process technology. I addressed the problem with him and offered to help; there was a lot of activity, but basically nothing changed.

Finally it was time for the two-by-four: I told him, "If you can't address this area, you can't remain in the position." That got his attention, and he came back shortly with a plan to reorganize slightly and bring in two new process-technology executives to take up the slack. It was a perfectly satisfactory solution. Charlie was left wondering out loud, "Why didn't I do this a long time ago?" Under pressure, he had forced himself out of his comfort zone. Since then, he has continued to face other comfort-zone issues on his own.

I measure a person's continuing growth in maturity by their ability to pry themselves out of their comfort zone, so they keep growing without having it forced upon them. In my opinion, prying *yourself* out of your comfort zone is a virtue to cultivate; in others, it's a virtue to watch for, nurture, and reward. Some of your players are going to be capable of rapid growth. Watch for this, cultivate it, and reward it.

SETTING GOALS

When I was at Bell Labs, no one from higher management ever came down and said, "These are your goals for the year." But they created an environment for growth by plunging us all in a sea of hardworking, brilliant people and letting us sink or swim. Everyone swam.

Bell Labs was an unusual situation, of course—one that most managers or leaders can't create.

The everyday approach is to get your people to decide for themselves to raise the goal. Your aim should be for your team to come to you and say, "Our goal on xx for the year is to improve by yy percent." The better job you can do at getting them to set the incremental achievable goals, the more likely they are to be motivated to achieve them.

As a manager you may, like a watchful coach, be sure that the bar goes up in achievable notches. Then again, you'll want them to set goals that are sufficiently ambitious. Stay alert and ready to intervene—making sure that the goals are accurately set to benefit both the company and the individuals.

TWO KINDS OF GOALS

I talked earlier about the two kinds of goals in business—continuous improvement and quantum change (sometimes referred to as "stretch goals"). Both are important.

In raising the bar, you will usually notch expectations up to higher levels one step at a time. If it's taking fifty man-years to design a new product, you set a goal of reducing this to forty-five, and, over time, progressively lower. If the average to complete processing the paperwork on a new order is thirty-five minutes, you set an initial goal of thirty-two minutes. Most goal-setting represents this kind of linear improvement.

Every now and then, however, you want to set a goal that truly stretches your people. An example is the effort at National that we call the "10X program." Instead of setting the reasonable but modest goal of reducing the time and cost of new-product development by 10 percent, we have asked, "How can we improve these by *a factor of ten*?"

Clearly a goal like this demands a far-reaching effort. One group at National, Data Management Design, set themselves a 10X goal of multiplying their revenues by a large factor within five years. To achieve this, they have identified six attractive new markets, developed strategies for each, and begun the creation of the systems, business processes, and other elements they will need, as just one aspect of achieving this difficult goal.

When trying to break the paradigms and significantly alter the way things are being done, I don't select a typical incremental change, but opt for radical stretch goals.

These must be used sparingly because they require a significant amount of the organization's energy. Incremental improvements, on the other hand, should be considered "business as usual." Whatever a result was last year, aim to do incrementally better this year. And be certain your people accept incremental improvements as part of the company routine.

Every time someone breaks an Olympic record, I think to myself that there is no limit to what we humans can do as long as we build up to it in an organized and sensible process.

HOW TO SET YOUR STANDARDS: ASKING THE COMPETITION

But how can you determine what level of performance is merely satisfactory, what is difficult but achievable, and what is superior? If it's always taken three weeks to close the company's books at the end of the fiscal year, does a goal of two and a half weeks represent a satisfactory level of raising the bar? Maybe other companies your size, in your industry, are taking four weeks, and your people are already achieving a stunning level of performance. Or maybe everybody else is taking two, and setting a goal of two and a half will still leave a serious problem unattended. Perhaps you're unknowingly tolerating lazy, sloppy behavior on the part of your accounting staff, or perhaps you're suffering from a lack of knowledge about current methods. Or perhaps you need new computers, new software, a larger staff. When you don't know what represents a reasonable level of performance, you need to find out.

The first stop is your trade association. They may have the answers you're after, or for a modest fee may do a study for you. But if they can't help, what then?

My answer will surprise many people: Ask the competition.

If you're the vice president of manufacturing, the corporate treasurer, or head of the accounting staff, pick up the phone and call your counterparts in several other companies.

The typical reaction I've heard to this suggestion from students, and from other managers and corporate leaders is, "Why would the competition tell you anything?"

Over the years I've found that as long as you're willing to share information—to give the other person the corresponding information for your company—you will generally get useful answers in return.

It's likely your counterpart will also see the information you're asking for as valuable: the questions you ask will probably trigger curiosity about how well the other company's operation compares with yours.

These contacts must be peer-to-peer: a VP of manufacturing isn't likely to share this kind of information with a manager, and the CEO of Company A may not even take a phone call from a VP of Company B.

This is not a project you can hand off to someone else. And it doesn't work well either if a high-level executive calls a lower-level manager: a VP of sales and marketing who calls product managers of other companies leaves himself open to being suspected of unscrupulous behavior—even by his own management.

If the initial question to your counterpart doesn't elicit a useful response, try making a more generous offer. You're going to be calling several companies and assembling the information into some kind of memo or report. Offer to share, not just some numbers about your company, but the complete results of your study. This represents highly useful information that any manager or executive should be glad to have.

Some years ago, I undertook a project like this in order to establish benchmarks on time to market. I called some twenty companies, offering each a copy of the results. Not one person refused.

People think it's hard to nail down comparative information, but it's actually quite easy. I've never had any trouble getting the answers that help me to determine what level of performance I can expect in my own company or operation.

LEARNING FROM WHAT WORKS

People learn by imitating what they see done right, but even better is learning from experience—surviving their mistakes and using their own common sense to avert a repetition. Confidence building is a primary benefit of the discoveries we make for ourselves rather than just following directives and rules.

There's a time for self-discovery and a time for following orders, but managers should not issue instructions that are too explicit just to save their own time and frustration. Part of transformation lies in encouraging managers to apply the wisdom of self-discovery as a day-to-day management tool.

I use the tool of self-discovery regularly. On one occasion the company was having a problem with low yields on some types of wafers. ("Yield" is the semiconductor industry measure for what proportion of chips are usable—low yield means high scrap, and high costs per good chip.)

The engineers and production people were adamant that they had tried everything under the sun, and the yields were as high as could be achieved with current technology.

Knowing that other companies were getting significantly higher yields with the same technology, I found myself gripping the arms of my chair to keep from throwing those statistics into the discussion. There are times when you have to bite your tongue because you know the answer, but you want your people to figure it out by themselves.

I quietly worked out a scenario with a friend—Al Stein, the chairman and CEO of neighboring chipmaker VLSI Technology, and then told my people that I had arranged for VLSI to run some tests for us. "But they won't be able to figure out what to do without some help and guidance," I said. "So we need to send over a team of key people to work with them."

The result was much as I had anticipated—the team came back from VLSI wide-eyed and all fired up, enthusiastically sharing with their coworkers the news of the success VLSI was achieving. Seeing it for themselves had swept away the mental blinder.

Raising standards now became a challenge that could be met and conquered. The knowledge that it was doable brought a new spark of enthusiasm to their work. People even tried to convince me of what could indeed be done.

The team leaders decided to reward and offer incentives to everyone involved in the improvement effort by presenting a very special-looking, custom-made coffee cup each time a new level of achievement was reached.

Some months later in an engineer's office, I saw a shelf with eight cups on it. Ever since, whenever I see that shelf of trophies, it reminds me again the value of finding ways for people to learn for themselves and motivate themselves.

MAKING A MISTAKE IS NOT A SIN

To build any organization up to very high standards requires that you allow people to make mistakes. I don't mean you should overlook mistakes, but if your people are neurotic over the possibility of making an error, they will be rendered incapable; they will seem to be busy but they will avoid challenging decisions and actions.

You want to become the kind of manager or leader who allows mistakes, but comes down hard on the failure to recognize, admit, and correct them. Making a mistake is usually of little importance; failing to learn from a mistake is a waste.

Your people must come to understand that a "high-performance organization" does not equate to being perfect. "Zero defects" is not a description of people, and "right the first time" only applies on the manufacturing floor—the ideal manufacturing process has no variation and creativity is absent by design. In manufacturing we aim for a "creative elimination of variation."

But that's not the governing principle for the rest of the organization. Especially when you are trying to transform your organization to higher standards, creativity is not only commendable, it's necessary for improvement. For some managers, this requires a 180 degree change in attitude.

In today's rapidly changing global business climate, unless you are improving, you are deteriorating. In other words, holding your own means losing ground.

To move forward, mistakes are inevitable.

PEOPLE WHO FAIL

What do you do about someone who, despite your encouragement and guidance, has proven over time unwilling to learn from mistakes and unable to grow? Okay, coach, when one of your team doesn't make it over the bar to the next height, what do you do?

Your first question should be, "Has this person reached his or her limit in this area or are other factors at work?" The problem might not be with the person, but with your coaching, or with some organizational constraints, or some cause beyond their control.

It is important to move precisely to assure yourself that the problem is not structural, that the person has indeed reached the limit of his or her ability in the current position.

Then remind yourself that although you must drive to high standards and transform to even higher standards, you must never lose sight of the humane side of problems. You will learn to draw conclusions like: This person is not going to make it over the high jump, but he might possibly handle the long jump or some other event."

Most people who are not succeeding know they are not succeeding, and suffer that reality even though they have not yet begun to face up to it. Those feelings of inadequacy contribute to depression and they often turn to complaining and blaming others. Uncertainty and an indecision that lasts for too long will cause high levels of distress—not just in the person but in those around them.

It's time for a change—not only for the organization's sake but in the long run for this person. The natural tendency, however, is to say, "Good ol' Joe has been with us for twenty years, great person, loyal to the company, works hard, shows up every day, needs the job, the rest of us can carry him." Does this make Joe feel better? No; it's patronizing. It's the worst thing you can do to his human dignity and his self-image. The uncertainty that hangs over Joe is like an invisible guillotine. But out of a sense of friendship and compassion, although it's the wrong thing to do, we don't face up to the situation with courage.

You need to find another position in your department, or in the next department, where the talents and capabilities of good ol' Joe will be a better fit—somewhere that he will still have room to grow. He deserves a chance to be successful again.

There will inevitably be some discomfort as you move him out of his comfort zone over to the new track. Whether he's grateful for the change or resentful, the chances are that when he begins feeling the glow of success once again, he's going to feel good about his new job and himself. A year later, he'll be back to thank you.

If you cannot place such people somewhere in your company, attempt to help them relocate to another company. Don't throw them away; help them find the right job. Ensure that a sincere effort is made to get them placed. One employee I helped get established in a job elsewhere came back years later to thank me for steering him into a great career in a totally different field.

Our own satisfaction is greater, not when someone else reaches out in support or kindness to us, but when we do the same kind of generous act for someone else. And remember that, as a leader, everyone is watching you and deciding whether this company-in-transformation is worthy of their continuing loyalty.

MAKING LAYOFFS UNNECESSARY

When every manager consistently follows this process of finding a more suitable position for any worker who is not meeting the new challenges, in time you find you have created an elite group of people holding jobs that match their capabilities. My goal is to reach a point where we no longer need to be concerned about layoffs because we have dealt with the problem at the source.

If National had been doing this through the 1980s, the company would never have needed to let people go. I believe that layoffs are a sign of management failure.

By moving the people who fail to meet your continually higher demands into better fitting positions, you create an organization that becomes lean and strong, increasingly better and better.

That's what happened at Rockwell. The first couple of years I was there we let some people go; they hated it, the other employees hated it and I hated it. I saw to it that we never needed to do it again. Occasionally someone would be eased into another job because they were not succeeding, or even, very rarely, let go for nonperformance, but we never had another sweeping reduction.

So one of the things you should be aiming for is this steady, continual refinement—selecting people through the process of seeing how they perform and helping them all you can. The aim is for everyone to be a winner, feel like a winner, and perform like a winner... and then, keep raising the level of their output.

FREEDOM AND RESPONSIBILITY

Management has always wanted more and more from their people, so there's nothing new about this. The difference is that as a manager or leader in this supposedly enlightened era, you not only want your people to keep raising their standards and output, you want them to do it because *they* want to.

You want them to be free to create, to feel validated, and also to think and work within the framework of the organization's purpose. Frankly, the old way of telling people what to do is much easier and faster, and there are many people tempted to run their organizations that way. But the world and the work force have changed.

Too often as leaders and managers we are faced with the trade-off between freedom and responsibility. There are many ways to balance these two values.

Most of your people can be given a generous amount of freedom. However, the benefits will only be realized if it is accompanied by absolute clarity as to responsibilities.

The historic wisdom that says you need to give very specific rules is mistaken. Too often managers micromanage—they're either unaware or can't seem to help it. The word micromanage is another one of those classy ways of saying that the person is still supervising and hasn't learned how to be a manager/leader. People who micromanage generally get precisely what they ask for, and no more. When given latitude, everyone can contribute; mail room people can come up with a more efficient way to distribute the mail than a route prescribed by a manager years earlier but still being used.

A simple and most basic goal of leadership is to get your people to manage and motivate themselves. Strive for and measure your performance by this difficult goal.

When your people decide to work on a Saturday, it should not be because you told them to, but because they're committed to the project.

What are the specifics behind the freedom/responsibility equation? Most important is that once you've agreed on what the expectations are, you must not arbitrarily change them. If you agree with a direct report that a certain task is going to be done by a certain date, he or she has the responsibility to see that it gets done as contracted. If not, you must be alerted in advance about the slippage and the person must be prepared to explain the cause and present the new schedule for your concurrence. You must be rigorous about requiring that people live up to these agreements.

As a manager you must be prepared to help get people out of difficult situations and to be understanding when the deliverable has turned out to be particularly difficult to achieve or even, due to special unforeseen reasons, impossible.

If you create a work climate like this, you're going to get a consistently higher level of results. Transformation happens when freedom combines effectively with responsibility.

NAVIGATING AROUND THE ROADBLOCKS TO HIGH STANDARDS

In the words of H. G. Wells, the English novelist and historian, "What on earth would a man do with himself if something didn't stand in his way." I suggest that you expect many things to stand in your way as you travel the unpaved road to high standards.

One set of problems arises when the objectives set by one department are inconsistent with objectives set by others. There must not be any white space between the departmental functions nor should there be any overlap, requiring managers to work horizontally as well as vertically—not just within their own organization, but also crossing the lines to work with other organizations of the company. To develop this skill, I bring in two managers, tell them what the problem is, and send them off to figure out who's going to do what. By now you know that I put a tight time limit on when I expect to know their plan. People will astound you with the creative solutions they are able to discover for themselves.

Sometimes problems occur when managers are not thinking horizontally. For example, at some point every department has to think about satisfying the customer. This is self-evidently important because a customer may have bought one particular product but also needs other products made by different divisions within your company. Unless the entire organization is responding in an integrated way, the customer will not perceive that you're delivering full value.

Solving the problem of synchronization of products and service at a consistent high quality level is essential for satisfying the customer. To transform the customer's perception of National has required improvement of the horizontal thinking at all levels; we're still working on it.

Consistency is a hidden agenda item at every meeting, and I am always on the lookout for any inconsistencies in theory. As you know by now, I suggest that you encourage disagreement and debate and individual thinking, but it can be a dangerous ditch in the road to transformation if you allow disagreements over basic values or goals. That's when you must assert strong guidance and bring everyone back on track.

DESIGNING FOR THE DESIRED BEHAVIOR

To meet the challenge of encouraging people to behave in the desired ways without giving explicit instructions, the place to start is by asking yourself: "What are the behaviors I want in my organization?"

And once you've decided on those behaviors, you need to ask, "What are the things I need to put in place so these behaviors will take hold? What environment do I need to create that will encourage behavior to grow in that direction?"

I heard about a woman who tries to strengthen her house plants by periodically turning them so the sunlight seems to strike them from a different direction. As these plants bend back toward the sun, they regain an elegant upright posture and seem to grow stronger.

If you want engineers, for example, to be highly creative, you need to show that you value creativity. Remember to remove from the environment the many bureaucratic obstacles that keep people away from their creative work. And forge an environment that regularly offers the engineers valuable exposure to outside lecturers, meetings, and so on that can enhance creativity.

You can achieve better corporate behaviors with the environment you create than you can with the movement of your lips.

KILLING PROJECTS

There's one uncomfortable behavior you must personally develop to support creativity and raise standards. You have to be prepared to kill a lot of projects. Just as nature throws off many variations and experiments that don't succeed, so must the transformation manager.

If what you want is an innovative organization that's trying a lot of new projects and processes, you must be able to *rapidly* make selections and decisions about what works and what doesn't. That's not at all obvious, and is frequently overlooked or misunderstood.

One of the skills you need to hone is this ability to kill off projects. Perhaps the word "kill" is inappropriate because you must do this with determination, not with anger. The truly valid decisions are never made when your jaw is set or your fists clenched.

How do you kill a project without stifling the creativity and enthusiasm of the people involved? The answer lies in something that has to be put in place from the beginning of the program: the definition of success, which must be part of the plan going in. If the team doesn't achieve what they agreed to, the "kill" decision won't come as any surprise.

Disappointment is inevitable, but I always require that the team do an "autopsy" and write a report on the reasons that the project didn't succeed. In this way, the company gains some benefit even from failed projects. Perhaps even more important, the team members have the sense that they have contributed some knowledge and learning to the company; this isn't much of a substitute for success, but at least leaves them with a sense that the organization has benefitted from their effort.

FOOTNOTE: A CAUTION AGAINST MISPLACED CREATIVITY

Question: Does the high intelligence level of workers always impact positively on high quality performance? Answer: The only place where intelligence comes before quality is in the dictionary.

Here's a typical example. The workers on the manufacturing line at the plant of a National subsidiary joint-venture in Israel were very bright and very creative. Though we've now sold most of our interest in the plant, when launching our operations there, we found that the employees were putting their talents and abilities into the work so thoroughly that each time a wafer came through, they would handle it a little differently.

Getting the operation to run smoothly was one of the toughest jobs I ever faced. In my frustration at the time, I referred to it as "the world's worst manufacturing operation." It damn near broke us.

The problems were only overcome when National established areas where creativity was prized and encouraged, and finally convinced the employees to be creative in those other areas. (I later found out that Intel's plant in the region was having a parallel experience.)

WRAP UP

The goal of continually raising standards does not lead to an organization that never makes a mistake, but to one that has a commitment to high standards, that tests and trains for continual improvement, and that respects and promotes individual talents.

It also leads to an organization that addresses and deals conscientiously with its personnel problems, an organization that tries a lot of things and rapidly makes decisions about what works.

An organization that has achieved high standards is one that can admit its mistakes and move on. To successfully transform your organization requires that you have the fortitude to kill off the failures so energies can be focused on trying new things. Don't aim desperately for being right; shoot for being successful.

Strive to create an environment that is not necessarily comfortable, but stimulating. In the process of measuring and benchmarking everything important, the manager and you and the whole world (it seems) will know how he or she is doing. And by putting the manager in a position to make real decisions, you will be raising the potential sense of accomplishment for the individual manager, and you will also definitely raise the stress level. This is a very desirable, productive form of stress, leading to what I call "the restless manager."

PART TWO—BUSINESS LEADERSHIP

SECTION 3

Fixing the Framework

11

Structuring for Success

"The modern company needs multi-dimensional managers who think and work in all organizational dimensions simultaneously."

Starting in about the late 1980s, a number of the National executives had been campaigning for greater decentralization. One who felt this strongly was Kirk Pond , who since 1988 had been running the company's Logic Division in South Portland, Maine. With National's money draining out at an alarming clip, holding tight reins on the decision making had appeared the logical route. What's more, it fit right in with Charlie Sporck's style of management.

"Charlie was functionally oriented," Pond says. "He believed in the highly matrixed organization—centralized manufacturing, centralized everything."

When the Board asked the executives for input on what qualities, capabilities, and talent they thought a new CEO should have, decentralization was one issue that kept coming up. So it struck home with the Board when during Gil's session, in his discussion of things he would do, decentralization was near the top of his list.

Business professor Bob Miles, who was involved from the beginning of the transformation at National, says "Gil decentralized the company structure very quickly. The GMs became the key financial entity."

The factories became semi-autonomous, self-functioning groups. In Pond's description, this was "not just empowerment, it was restructuring."

The years at Fairchild Camera and Instrument Corporation provided the crucible for Gil's concepts of management. As Gil's self-appointed mentor, R&D director Jim Early offered guidance not only in the technology areas but also in the people-aspect of management.

Typical of Early's style was his eagerness to help Gil acquire a mentor on the business side, as well. After Gil had been on the firing line for four years, developing his management skills by learning to handle the reins on a team of undisciplined engineers, Early decided he was ready to be moved out of an R&D environment into mainstream management. From Tom Longo, one of Fairchild's senior vice-presidents, Early obtained a commitment allowing Gil to sit in on Longo's weekly management review meetings.

Gil says—

> *From Tom I also learned how not to handle people—his style contradicted what I had learned from Drucker and seen from Jim Early.*

Out of this comes a clear distinction that Gil now makes between the hard side and the people side of managing, while recognizing an inherent challenge to be effective in both areas. "Like most other high-tech managers, I tend to over-analyze things," he says. "I place too much emphasis on the analytical side and have to remind myself regularly to stay involved in the human side."

COMMAND AND CONTROL

Years ago (this will come as a surprise to many) National Semiconductor was a larger company than Intel. When Intel proved highly successful with a new style of chip that came to be called the "microprocessor" (which has since, of course, become the brains of the desktop computer and computer-based products), it made sense for National to go after a share of this burgeoning market.

The effort was never successful, because National's approach was to design and market chips that seemed to hold promise, while Intel was doing a far better job of finding out what customers wanted, anticipating changes in technology, and keeping in touch with global factors.

National's microprocesssor program turned into a $100 million black hole that had dire consequences for the company. Yet there were many people within the organization who saw the problem, and who may have had solutions or, at least, may have known where to look for them. But in an organization with a rigid, traditional structure, these people had no way of influencing the course of the company or even, in most cases, the course of their department or their product.

National's experience with microprocessors is by no means an isolated situation; most companies have similar horror stories, and if you've been in the business world very long, you likely have a few that you've observed first hand.

The underlying problem is with the organizational structure that has been standard in companies since the dawn of the industrial age and before, one that parallels the structure of an army—a style we've come to refer to as "command and control."

Command and control was an appropriate, effective method for business during an era when time wasn't so critical for success and the rate of change wasn't at such a breakneck pace. It was also more suited to an era when people were raised and educated to respond to orders. In days when companies were not as dependent on inventive minds, creative solutions, and competitive advantages, command and control worked well enough.

The command and control structure has a preferred axis like a stiff spine running up and down the organization. You do what your boss says, he does what his boss says, and the managers make it very clear: Don't be creative, just do what you're told.

Although command and control is rigid and unnecessarily restrictive, there is a time when it's justified. In a crisis situation, command and control is often the only appropriate choice, the only way to proceed if you're going to prevent a disaster.

But that's the exception. More normal conditions demand a more lenient and more fruitful style. Think about a common business activity like preparing a five-year plan; you want to synthesize the creative thinking of the entire organization, and command and control severely stifles the kind of contributions needed to enrich the plan with the best thinking from throughout the company.

But if not command and control, then what?

THE CUBICAL HYPERSPACE

The typically buttoned-up corporate structure is often represented as a pyramid—with the chief executive at the pinnacle and the reporting levels forming neat, orderly, widening layers. This is the paradigm of the command and control organization.

As I thought about the problem of taking a traditional corporation with its pyramidal structure, and shaking off the strictures—molding it into a more vibrant and fluid organization that unleashes the workers to greater freedom of contribution—I began to conceive a corporate structure for which the graphical representation is a three-dimensional object, a cube, in which each dimension represents a different way of thinking about the company's activities as illustrated by this drawing—

(In reality, the "cube" should have more than three dimensions making it a "hypercube" but I don't know how to draw a picture of that!)

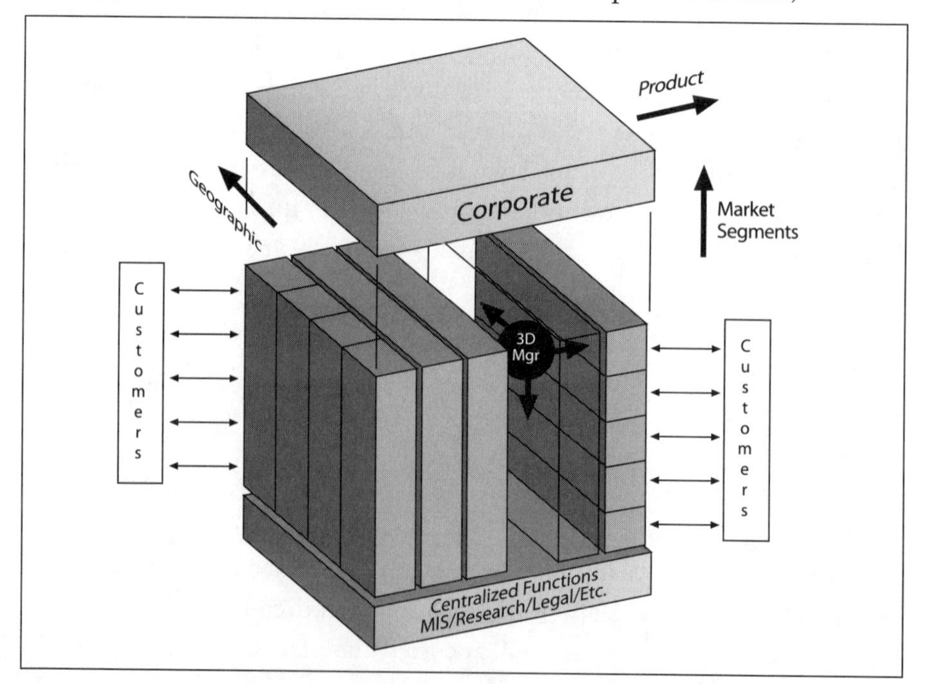

The Three-Dimensional Corporate Structure, or "Cubical Hyperspace"

This way of viewing the business allows us to focus on two quite vital issues that are otherwise generally overlooked: how customers want to interact with our organization; and how we get people within the organization to throw off the shackles of command and control, and begin thinking and acting in new, more effective ways.

1) How the Customer Wants to See Your Company

Organizational structure is not something that has impact only inside the company; your customers and others who deal with you are affected by it in ways that are not always obvious.

Corporate divisions tend to be organized in terms of related technologies or functions. Factories, to be efficient, have to build more or less similar products. And over time it's natural that we have extended this concept to the organizing of other activities like marketing, sales, public relations, HR, legal, and the rest.

Take a different perspective and look at this organizational tradition from the viewpoint of the customer. When General Motors places orders with National Semiconductor, they want analog chips, digital chips, and other products. They really don't care how we're organized internally—that each of the products on their shopping list may come from a different division. When they call to place an order, get technical information, or find help solving a problem, they don't want to be told, "We only deal with the White product line; for Black you'll have to call another department."

They don't want to think about the internal intricacies of National Semiconductor. And your customers don't want to think about the internal arrangement of your business, either.

As you begin to succeed in convincing managers to think of their jobs in terms of the cubical hyperspace depicted in the diagram, they become able to conceive their role not just as satisfying vertical demands from above, but in terms of addressing other dimensions simultaneously... which, among other benefits, leads to new attitudes about satisfying the customer.

2) No Preferred Axis

The second powerful advantage of encouraging the kind of organization structure represented by the cube diagram lies in the power you gain when you achieve a more fluid organization, one in which people are liberated to view the company in multiple perspectives and think about it in new ways.

A fluid organization is one capable of reorganizing *every day*, if necessary, to serve the needs of the marketplace. When a competitor comes to market with a new product, when new technology skews the market focus, when geopolitical shifts change the market climate in some foreign country or region—the fluid organization will be able to respond and adjust quickly.

WORKING IN THE CUBICAL HYPERSPACE

You must find the ways for everyone to become equally agile at working along any of the axes of the cube diagram—reducing the decision-making that takes place along a "preferred" axis. Ideally, no one will operate exclusively or primarily along a single axis.

That's a very important notion, worth repeating; people must not operate along a single preferred axis. You are not going to achieve a transformed level of decision-making if you have an organization where one structure—let's say, the product-line manager—is preferred over the marketing people or the engineers or some other group in the organization. You'll merely get what the product-line managers want to do, and although they may be making adequate decisions, most of the time you won't be getting the best decisions possible.

The multi-dimensional structure has another great virtue: it provides the kind of flexibility that allows people, even encourages people, to step forward and make their voice heard when they recognize a problem or see a previously unnoticed opportunity. At National, a manager at any level can organize an ad-hoc team to address an issue (a process described in detail in Chapter 15.)

THE MULTI-DIMENSIONAL MANAGER

For any *large* company, however, the situation is complicated. On the one hand, you have to organize the product, development, and manufacturing around some core of technology and specific applications. But this often leaves you unable to address the questions that can only be seen in one of the other perspectives represented by the cube.

Managers who successfully conceive of their role in a view that reflects the multi-dimensional structure are people who will be able to function as "multi-dimensional managers," which is what you're hoping for. (On the three-dimensional chart, we show this as a three-dimensional capability, but it's the "many"-dimensional you hope for in real life.) On any given day, such managers can look at their operations as the sum of the business they did in Europe, the business they did in Asia, etc. At the same time, they can view their operation as the sum of the business done by each of their product lines, and so on.

Rather than the myopic decisions encouraged by "preferentials," the modern company needs decisions from managers who think and operate in all dimensions simultaneously.

STRUCTURE AT NATIONAL SEMICONDUCTOR

Now take a look at how the cube model for organizational structure has been applied at National Semiconductor (circa 1994).

Reading across horizontally, activities are sliced in terms of corporate groups and product lines—Analog, Data Management, and so on. The vertical dimension represents the company in terms of market segments—automotive, military/aerospace, communications, etc. Going into the page is a representation of our geographical entities—the Americas, Europe, Japan, Southeast Asia—the way we look at our business when we talk about sales or markets in a particular region.

This structure, installed at National in 1993, was based on the insight that the smoothest way to throw off the chains of command and control is by creating decentralized organizations—smaller groups that address particular areas.

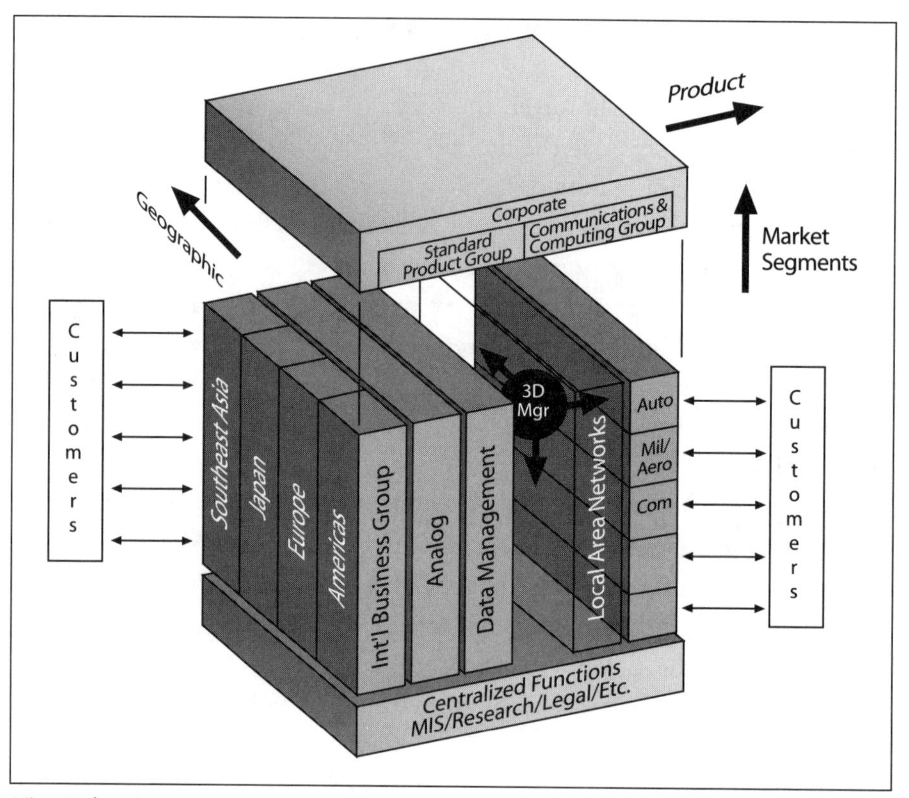

The Cubical Hyperspace at National Semiconductor.

THE KEY TO DECENTRALIZING: GROUPS THAT COMMUNICATE

The idea of decentralization has been around for a long time; what's new here? One chief problem with a decentralized structure has always been in the communication between the groups. Or rather, the lack of it—in some companies, communication among these independent suborganizations hardly exists. This is a reason why many companies that decentralize decide after a few years that the new structure is worse than the old one, and revert to something akin to what they looked like before.

National's answer was to put into place a number of integration functions that cut horizontally across the company and provide the missing lines of communication.

One illustration to clarify: National's R&D is carried out by a group we call the Central Technology Organization, headed by chief technology officer Charlie Carinalli. Think of this as his vertical responsibility within the same kind of a traditional and familiar structure that other companies use.

But at the same time, Charlie has another responsibility, a horizontal one: he is "Engineering Leader" for the entire company—a functional job under which he is accountable to me for the excellence of engineering work throughout the entire company. In this role he needs to be in contact with groups that do not report to him, and he is challenged to address engineering weakness wherever he finds them. He goes from division to division and operation to operation, assessing the quality of each engineering effort. Where he finds weaknesses, he works with the people toward improvement.

In the process, of course, he inevitably keeps in touch with who's doing what, whether two different groups are working on the same issues, what roadblock one group is encountering that another group may be able to help with, and so on.

Note that this is not a policeman's role—he's not looking over people's shoulders to whip them along or judge their performance, but, principally, to facilitate horizontal communications. The result is a significantly higher degree of collaborative, cooperative, coordinated work across the company.

Every senior manager at National now has a horizontal job as well as a vertical one. Pat Brockett is responsible for the marketing competence of the whole company, Randy Parker for quality and reliability, and so on—creating linkages across the company.

Without this horizontal effort, we would no better than twenty small companies. In some ways worse, because twenty small companies could move faster.

Five or six years after starting this effort, I expect to reach the point at which we'll no longer need the horizontal roles, because our employees by then will take for granted that horizontal communications are acceptable and expected. We have already gained appreciable ground in that direction—most of this communication now takes place with no management effort needed.

TOWARD A FLEXIBLE STRUCTURE AT NATIONAL

The product-line structure—which is where much of the decision-making and design, and all of the manufacturing, generally takes place in a manufacturing company—has been quite orthodox at National. Apart from the international business groups, the product lines report to divisions, which are organized into two groups (until mid-1995)—Communications and Computing, under Rich Beyer, and Standard Products, under Tom Odell.

In 1993, we created an additional structure that we call "Strategic Market Segments" (SMSs), to make us more effective in addressing large market segments—such as automotive, personal systems, or business communications. These segments may overlap several product lines, which may be in different divisions or even in different groups.

The SMSs are an attempt to answer the question of how we maintain the internal competence that resides in a product line, and at the same time become more fully responsive to large market segments. By creating the SMSs, we have created greater fluidity. These might be called "virtual overlays," since they are not separate organizations but use the same people, wearing different hats.

As part of the SMS effort, and to move people toward being able to function as three-dimensional managers, we have also created general managers for each geography—the Americas, Europe, Japan, Southeast Asia—and we've given them equal clout with the product lines. For example, Hans Rohrer, who runs our European operations, has as much to say as the head of the logic group about what we do with logic products in Europe. They are, if you will, business partners with equal votes, who have to decide together on approaches and programs that makes sense from both their perspectives. Neither one has a trump over the other.

(This type of relationship has in the past been frowned on. The old rule was that each person had a label on their forehead for a specific assignment, and shared responsibility was to be avoided at all costs. But joint-responsibility arrangements are becoming more common, and for good reason.)

BALANCING ACT: STRUCTURE WITH FLEXIBILITY

The structure of an organization should provide a roadmap and a guide for getting things done. In a company where people literally don't know how to get something approved, how to purchase a new piece of equipment, how to hire a supplier, a great deal of time gets wasted. While this may seem like a situation that simply couldn't exist, the reality suggests otherwise.

Typical is the angry e-mail I recently saw quoted, that had passed between two managers in a high-tech company. The gist was, "You can't hire a vendor the way you just did for the vice president's speech. A purchase order has to be issued by me before a vendor begins, and if you try doing it that way again in the future, I won't do anything to help you get the person paid." In other words, you don't know the structure that has been created for this function, and I won't help you until you understand the structure and follow the procedures.

In the rigid structure that we refer to as bureaucracy, most people seem dedicated to following the rules rather than helping to achieve the Vision or the goals.

The opposite extreme of the overly rigid structure is what you see in many start-up operations. Some companies become successful so quickly that they never take the time or recognize the need to get effectively organized. Fast-growing start-ups usually refer everything back to the entrepreneur, the founder. Ask what someone does and you'll get a vague response; ask what the procedure is for getting a new project funded, and you'll likely be told that "it depends"; press a little harder and you'll be told that the boss has to give his okay.

Neither the overly rigid style nor the seat-of-the-pants form of the startup is the structure you need to keep the wheels turning smoothly in a modern organization. If there are organizational traffic cops on every corner, they will keep your people intimidated. Traffic will jam and people will keep their cars in their garages. They will be safe and you will be sorry.

Instead, you need attendants at every intersection and attractive street signs to provide guidance and direction. That's an organizational model to clarify directions and provide support rather than inhibit the progress that achieves goals.

The goal is to have enough structure to function, yet not so much that your organization gets in the way of your goals.

STRUCTURE, PRODUCTIVITY, AND SPAN OF CONTROL

Peter Drucker has written that "productivity is the true competitive advantage." Is productivity really a fundamental issue, or something way down the line of priorities? To put this another way, is the productivity of the most successful companies consistently better than not-so-good companies?

In the early 1980s, the Chicago-based international consulting firm A. T. Kearney, did a study on "High Performing Companies," comparing the top-performing companies in the Fortune 500 crowd to the average for all Fortune 500 companies. Among other factors, they looked at "span of control"—a figure representing how many people, on average, report to each manager. What do you think they found?

In top-performing companies, the average span of control was 7.5; in the Fortune 500 as a whole, it was 4.5.

I believe that what's happening here is this: When the span of control is too narrow, bright managers have lots of excess energy and under-used talent, and are tempted to justify their position and salary by doing more than they need to do—overcontrolling and micro-managing their subordinates. But given a sufficiently wide span of control, they can only survive if they learn how to lead and empower.

PepsiCo found this out by experience. Wayne Callaway, the company's chairman and CEO, told me the story of some managers in their Pizza Hut chain who were overbearing. He solved the problem by increasing their span of control from the five or ten stores each manager had been responsible for, to *forty*. At that point the challenge became lead and empower, or sink.

All of these managers kept their heads above water. They survived and thrived, and, I'd bet, found greater satisfaction. As for Callaway, his satisfaction is clear from the way he tells the story.

WRAP UP

The familiar command and control structure, the rigid, layered hierarchy in which each person receives instructions from above and gives instructions to those below, is widely recognized as no longer viable. Part of the reason that many companies continue for years with a bad decision or unprofitable line is that a command and control structure stifles employees who have an idea or an insight that goes against accepted company dogma.

The solution at National has been to move toward a new structure, graphically depicted as a cube, in which we challenge our employees to operate, not just in the up-and-down direction of command and control, but in a number of dimensions simultaneously. In the example used in this chapter, the three dimensions are product, market segment, and geographic. We refer to the new type of worker who operates within this structure as a "three-dimensional manager" or a "multi-dimensional manager."

This multidimensional structure also helps people to start thinking about customer needs in a new light—coming to understand that the way in which a customer views the company, and wants to interact with it, is entirely different from the way the company looks to those on the inside. Because customers see you one way and people inside the organization see you another, managers need to recognize that they stand "at the crossroads," requiring them to understand the customer's perspective.

During the transition period to the new corporate structure, senior National managers have been given additional, horizontal, responsibilities across the entire company (for research, or quality, or marketing, etc.). The goal here is to overcome one of the great barriers to the success of decentralization, by providing effective communication horizontally among the discrete units.

Another important route to improving corporate structure comes by increasing the span of control, making each manager responsible for authority over more resources or programs. This forces managers to lead rather than micro-manage; it can also, over time, appreciably reduce the number of management layers in the company.

The challenge is to be thoughtful enough and perceptive enough to get the result you want by creating a system that encourages people naturally in the desired direction. It is up to the leader to create a structure that supports this goal.

12

Organizational Excellence—Elevating Performance and Achievement

"A company that isn't pursuing the hard Ss is probably not going to survive. A company that isn't pursuing the soft Ss is probably not going to ***win.****"*

As the transformation continued, new managers and executives joining National needed somehow to be brought quickly up to speed in the philosophies that were taking root and that were largely responsible for the improvements. Helen Peters, who arrived in April '94 as a director in Human Resources, was sent to spend a week of training in Monterey, California, with a group of other new directors and vp's.

The attitudes were typical in a situation like this. Most of the people were "reluctant and facing the week with more than a little cynicism," she recalls. "As seasoned managers who had been hired from the outside, we viewed ourselves as part of National's future, not National's past. Perhaps employees who had been in the organization for a long time needed this, but certainly not us. We were different."

By Tuesday afternoon the situation hadn't improved, and instead was getting worse:

> *The breakout group I was assigned to came to an impasse. There was a major disagreement between two of the members, who sulked in chairs on opposite sides of the room while the rest of us put together our material for a presentation defining what National's Vision meant to us.*

> *In an attempt at reconciliation, we asked our warring team members to be our spokesmen; one of them, Sandy Sanderson, reluctantly acquiesced.*
>
> *We were apprehensive about exactly what Sandy would say. When the moment came, instead of speaking to the flip-chart we had prepared, he proceeded to make an impassioned speech about the Vision. He talked about the need for the Vision to become a personal one; about how National was on the precipice of a major paradigm change; and how, to make this shift and achieve the Vision, we would all need to 'expect the best' of ourselves and others.*
>
> *At the conclusion there was a moment of silence, soon replaced with a loud burst of applause. In that moment we moved from being a collection of individuals thrown together by coincidence, to a team with a shared sense of purpose.*
>
> *Of course, when we presented our ideas to Gil Amelio at the end of the week, Sandy was our lead spokesman.*

Out of this experience, Helen Peters believes, "We shared a sense of responsibility and partnership in the building of the new National."

★★★★★

When Gil Amelio, having taken over the SPD division of Rockwell, invited Emory business professor Bob Miles to help straighten out the problems, Miles' curiosity was piqued. Despite being warned off by a friend because the division was thought to be unsalvageable, Miles met with Gil and decided he was "truly committed to the idea of being a transformation leader."

It was quickly clear to him that Gil "was different from the hard-headed engineering managers at Rockwell—he liked to communicate, and understood its importance."

Miles also found that Gil was "asking the right questions, and putting emphasis on process," in contrast to most engineers and engineers-turned-managers, who, Miles believes, "generally prefer the specifics of a blueprint to the messiness of a process."

THE ORGANIZATIONAL EXCELLENCE PROCESS

Some people think the quest for "excellence" was a corporate fad that had its day and has now faded. There may be a kernel of truth here—but only in the sense that the word has been so overused, it's been rendered almost devoid of meaning, the way West Coast beach youngsters use "dude," "rad" or "y'know."

Excellence is as important as ever, and an essential for transformation—but, like virtually everything else in business, the quest must be part of a well defined program or process.

The Organizational Excellence program we now use at National has evolved from the version Professor Bob Miles brought into Rockwell and developed with me at the SPD division. It aims to elevate the organization's performance and achievement to the highest possible level.

So that we begin this on the same footing, here is our definition—

Organizational Excellence

> A management process that aims to align effort and provide focus by using participatory, team-based consensus within an established framework.

Once a course for transformation has been charted, you need a pragmatic process for making employees highly effective—providing a mechanism for change, and enhancing their energy level, motivation, and commitment. At National, Organizational Excellence affords that process.

TQM, OE, OR BOTH?

But of course, there's more than one road to excellence. Aficionados of management theory recognized that the world takes sides over different approaches to the subject. Of these, most people gather around either of two campfires.

One comes at the subject from the quality side; this is the approach of Total Quality Management, and any number of similar programs under other names. The advocates of TQM would maintain that all the methods

I use—everything in this book—fits very neatly into the framework of TQM. They would just see it as getting started from a different initial point of reference.

The other leading approach comes at the same subject from the viewpoint of a program like Organizational Excellence.

Probably by the end of the day, the two groups, comparing notes, would find that they had covered pretty nearly all the same issues and agreed on nearly all the approaches. In the end it does not matter whether you favor OE or TQM as a basic management philosophy: a fervent application of either will move you in the direction you're seeking to travel.

Having said that, I want to share my personal views on the subject.

While I strongly support the TQM program at National, at the same time I have reservations about the underlying assumptions of the approach. TQM does not, in my view, articulate clearly enough the various responsibilities and roles at the different levels up and down the organization.

Among the most challenging hurdles facing a manager is learning to walk the balance beam between the reality side of management and the emotional side. I find that most TQM material has paid very little attention to the emotional part.

When I started examining the options years ago, quality management philosophy was on the level of Phil Crosby's *Quality is Free*,[1] which included a bare minimum of people orientation. My sense even back then was, "This approach by itself is not going to help me transform my company—not in this era!"

Programs like TQM change over time as different contributors reinterpret and expand the earlier version; the elements I find missing may eventually be added.

From another perspective, people often look at my OE approach, shrug their shoulders and decide, "This won't work—it doesn't mention quality." Yet quality is in there, all right—but approached differently, from within a different, and in my opinion better, framework.

Experience teaches that change only happens in an organization when pressure is applied from different directions simultaneously. One way to do that—the way I prefer—is to use TQM to apply pressure from the bottom, while Organization Excellence is used to apply pressure from the top.

I start the OE process with the senior managers and work the program down through the organization. Within the same time frame, TQM is launched at the grass-root level and percolates its way up. Ultimately all levels are appropriately covered and improved.

LAUNCHING OE AT NATIONAL

Our "Executive Off-sites—the quarterly, off-campus gatherings that I use for my top forty or so executives—have been the framework for launching Organizational Excellence at National.

One of my principal reasons for these off-sites is to educate National's corporate leaders in the management and decision-making skills that experience has taught me are necessary. While much of our effort focuses on catching the team up on developments in each part of the company, and on hammering out issues about steps to be taken, we also spend perhaps one-third of our time together in what some of the executives probably call the "Amelio as professor" mode.

Most of these women and men haven't sat in a classroom for a long time. But transformation is about attaining and absorbing new knowledge and perspectives. Continued expansion of information and ideas is essential—and that doesn't mean just for middle managers. The example for intellectual openness must be shaped at the very top.

THE ELEMENTS OF OE

What is it you really aim to achieve with Organizational Excellence?

Envision an organization where every manager, given a challenge, would make the same decision the organization's leader would make. Most leaders would agree there is one characteristic that must always exist in the ideal organization: it must provide a way of focusing and amplifying the collective wisdom of its people.

Is there some possibility this fantasy could become reality?

In my view, "amplifying yourself" or "amplifying the effectiveness of the talented individuals in an organization" is a fine way to describe the ultimate goal of Organizational Excellence.

I have a checklist of four things I believe you must do to make that happen—

The Basic Elements of Organizational Excellence

Direction Setting and Visioning

The process must begin with the defining of a Vision and the setting of direction (Chapters 1 and 3). Every manager within the company must know where you are headed as an organization and feel that they have been a part of articulating the Vision and direction.

Eliminating Obstacles

Every organization that is more than a few years old and has more than a handful of employees has likely planted more procedures, rules and requirements than it needs or should have. The problem lies in filtering the valuable ones from the other kind, the bureaucratic roadblocks that keep employees from succeeding.

Educating

As an on-going part of the transformation process, employees need to acquire the skills and knowledge that the process will demand. Every manager must be committed to enhancing their own and their employees' level of competency and currency.

Informing

> Keeping all employees informed seems obvious—but try asking people in your organization whether they have ready access to the information they most need, in the format that's most useful to them. Employees at every level continue to report feeling left out of the information loop. The ideal would be if each employee could press a key on his computer, and see a presentation of exactly the information he or she most needs, in the desired format and arrangement. Every manager must have access to accurate, usefully presented, and timely information about what's going on in their department, as well as in the other parts of the company they interact with.

When these four basics are in place and moving ahead, I can just about guarantee that the manager who's running your plant in South Bend or South Africa, assuming they've fundamentally competent, will, when they can't get you on the phone, 99.9 percent of the time make exactly the same decision you would make.

THE 7S MODEL

Organizational Excellence provides a structure, but some kind of framework is also necessary for the action to take place. Any one of a number of frameworks are suitable for this purpose; at National, the one we use is the 7S model, which provides a process for achieving fundamental change in the way people go about doing their tasks.

The 7S model, in case you have forgotten or haven't encountered it before, calls on managers and companies to organize their efforts around these seven elements—

- strategy
- systems
- structure
- staff
- style

- skills
- shared Values

This framework, which was conceived by McKinsey and Company in the late 1970s, came to my attention about the time I was working on the transformation at Rockwell, and proved to be one of the most effective approaches I have used.

It's informative to remember the origins of this model. Here's how it's described by Tom Peters and Robert Waterman Jr. in their book, *In Search of Excellence*[2]—

> Early in 1977, a general concern with the problems of management effectiveness, and a particular concern with the nature of the relationship between strategy, structure, and management effectiveness, led us to assemble two internal task forces at McKinsey & Company. One was to review our thinking on strategy, and the other was to go back to the drawing board on organizational effectiveness... (p1)

In their study, the McKinsey people looked at middle-sized companies and asked, "What are the attributes that divide companies at the $100 million level, from companies at the $1 billion level?" Peters and Waterman report the results—

> Our research told us that any intelligent approach to organizing had to encompass, and treat as interdependent, at least seven variables: structure, strategy, people, management style, systems and procedures, guiding concepts and shared values (i.e., culture), and the present and hoped-for corporate strengths or skills. We defined this idea more precisely and elaborated what came to be known as the McKinsey 7-S Framework. With a bit of stretching, cutting, and fitting, we made all seven variables start with the letter S and invented a logo to go with it. Anthony Athos at the Harvard Business School gave us the courage to do it that way, urging that without the memory hooks provided by alliteration, our stuff was just too hard to explain, too easily forgettable.

> Hokey as the alliteration first seemed, four years' experience throughout the world has borne out our hunch that the framework would help immeasurably in forcing explicit thought about not only the hardware—strategy and structure—but also about the software of organization—style, systems, staff (people), skills, and shared values. (p 9-11)

Their depiction of the 7S model remains virtually the same some fifteen years later, as shown in the figure. The authors comment that "some of our waggish colleagues have come to call (it) the happy atom."

Although Tom Peters was one of the people who originally created the 7S Framework, he has more recently belittled the concept on the grounds that it presents a static model of a business, whereas any business is really a dynamic entity.

This misses the point. As a jet pilot, whenever I get in my airplane, I run through a checklist. I'm sure I know the checklist by heart, but no matter—I would never, ever, take-off without using the written one. Without the checklist, I might forget one switch that would keep me from successfully and safely reaching my destination.

That's what the 7S Framework provides—a manager's checklist to be sure that all of the essentials are thought through and all the switches are in the right positions.

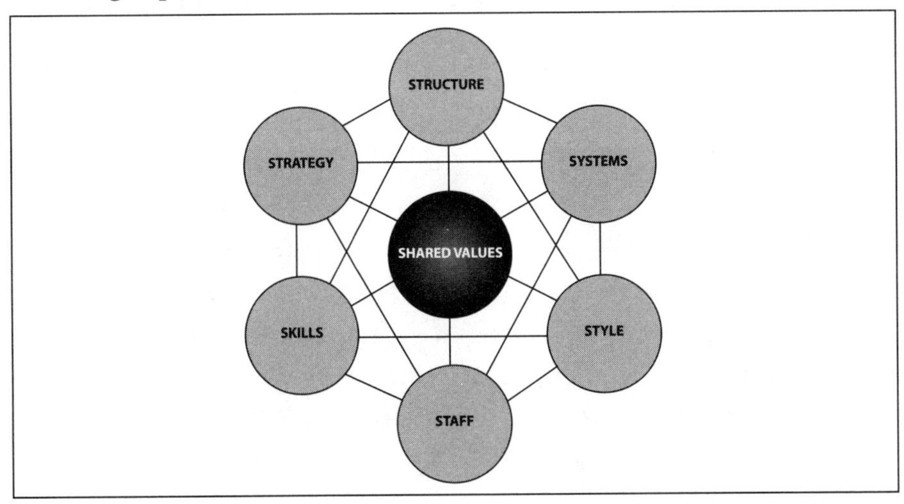

The original 7S diagram, from ***In Search of Excellence***.

USING THE 7S FRAMEWORK AT NATIONAL

In applying the 7Ss at National, we began by asking—"Where are we along each of these measures? What's our strategy? What's our style?..." And we also examined how the 7Ss would interact with each other, and whether they were self-consistent.

We then asked, "What do we want to do differently?" We developed a definition for each of the seven issues, and worked out the goals and objectives to go along with it. In this way, the 7Ss played an important role in helping us define and launch transformation.

Once the process was well under way, the 7Ss became a tool we called upon when it was needed, rather than something referred to on a daily basis. If we're about to make a change, we go through the steps much as we did initially—"How do we stack up on the 7S list? Where are we today? What kind of systems do we have, what kind of structure, what's our style?" The answers to these questions guide us on how we need to design the changes.

As a check list, the 7S Framework is a way of insuring that you don't overlook any important elements in the specific change effort you're addressing. I've learned that when you leave out even one element, the change chemistry doesn't happen. I want us as careful in our transformation lab as a pharmaceutical company is in their biochemistry lab. We leave nothing to chance and have check lists galore.

WHAT IN THE WORLD ARE THE "SOFT Ss" AND WHY ARE THEY IMPORTANT?

Everyone who works for some time with the 7S Framework begins to recognize that the seven items on the list fall into two separate categories. Strategy, systems and structure are concrete, quantative, left brain efforts, and the shorthand we use to describe them is "the hard Ss."[3] By contrast, the other four—skills, staff, style, and shared values—tend to be more qualitative in their dimension.

Skills asks, What are the predominant competencies and capabilities for the company in order to make us into what we want to become? At

National, among the skills we need are those that contribute to a competency in "analog and mixed signal" technologies, which are where we have historically excelled. In contrast, because Microsoft is essentially a software company, the ability to conceptualize very abstract problems is a critical skill.

Staff raises questions about whether we value people who are free thinkers, the individualists, or do we seek out people who are "by the book" and always do things the same way. Staff has to do not just with the kinds of people you have, but also with how people come together in communities of practice, and how they interlock with one another.

Shared values has to do with the culture of the company. For many years IBM was reflected in a culture of dark suits and plain accessories. For Apple Computer it was jeans and the every-Friday beer bash; in the early days, even bare feet were tolerated as acceptable business attire! Southwest Airlines takes no seat reservations and provides no food, but encourages its crew to improvise humor into their announcements to passengers.

These soft S, people-based issues are radically different from company to company; we can't write specific formulas and prescriptions for them.

The McKinsey study concluded that the human factor is the essential ingredient separating the truly great business performers from those companies that were just making it. In order to make a difference, the soft skills must be clearly defined, understood, and communicated.

A company that isn't pursuing the hard Ss probably isn't going to survive. But a company that isn't pursuing the soft Ss is probably not going to *win*.

Out of this came the notion of trying to teach American management something that they have historically never been very comfortable with—the soft skills. A manager didn't get to the board room talking about soft skills, and definitely didn't talk about them once there. But the reality was that people in the best companies have been paying attention to the soft Ss all along.

You need to be, as well.

THE ORGANIZATIONAL EXCELLENCE ROADMAP

Some of the changing priorities that occur when you carry out the OE process are depicted on what we call the Organizational Excellence Roadmap. The roadmap can be thought of as one way of describing the management agenda—what you're trying to do in the transformation.

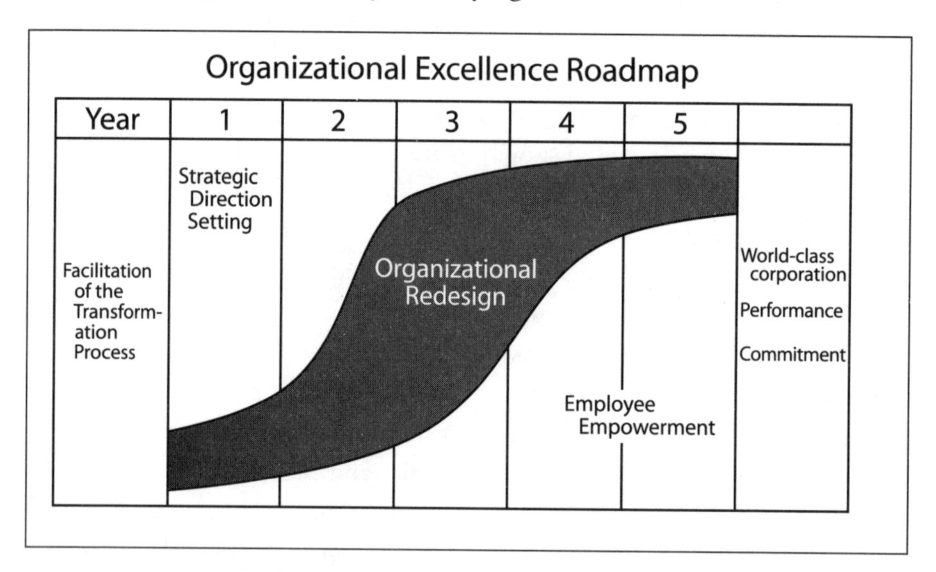

Organizational Excellence Roadmap.

At the start of transformation, the leader of the organization provides the strategic direction setting—represented by the white area at the top of the chart. Employee empowerment is a goal, but probably exists in only very limited areas. Management tasks in the early phase are aimed at encouraging empowerment.

As organization redesign starts to take hold, employees begin to understand the idea of empowerment, and begin to take the reins more frequently. Decision-making begins to be successfully pushed down the line.

Ultimately the organization's leadership can confine their strategic direction setting to the very highest levels.

I find this roadmap useful as a way of reminding people that executives and managers need to manage in ways that help make empowerment the accepted practice. The progress represented by the chart doesn't just happen; it only comes about as the result of a carefully orchestrated process.

We also show on the Roadmap where this process takes you: to become a world-class corporation, exemplifying high performance and high commitment.

OE: ALL AT ONCE

If you buy a sailboat that's been lying on the beach unattended for several years, you can't go to sea in it while you methodically and patiently carry out the needed repairs one at a time; you'd likely drown before you got to the second item on the list.

The same with Organizational Excellence: if you concentrate on fixing one problem at a time, you might be out of business before you've made much progress on the repair list.

The 7S framework carries a message that, whether you like it or not, you've got to juggle all of the 7Ss simultaneously. You can't do one at a time or you'll never get the business fixed. That's one of the most important lessons of 7S, and one of the reasons we use this approach: it helps us remember to make sure we've got balanced goals in each of these important categories.

SEEK GOOD HEALTH, NOT A CRASH DIET

To shift metaphors, a company in trouble faces the same problem as an overweight person: a crash diet may get the weight off, but he may not be healthy afterward.

It used to be a joke in Hollywood that when a studio's pictures weren't doing well at the box office, they would cut expenses by firing the secretaries. It's no joke that in many companies today, when the market goes soft, the boss tells everyone, "Cut 5 percent across the board." Now the company has to do the same work, but with 5 percent fewer people, 5 percent fewer resources, 5 percent lower budgets. After a period of this, the company looks better but is almost certainly in a significantly weaker state of health.

Remember that your ultimate goal is transformation, not turnaround—health, not misleading appearances.

So Organization Excellence needs to be looked on as a process, not as a short-term fix. Pursued consistently, it will continually reveal weaknesses in the organization.

Once a generation, every mid- to large-size company needs to undergo a major transition. The market forces it on you. Here, too, Organizational Excellence and the 7S model can help you succeed.

OE FOR YOUR ORGANIZATION

The same kinds of Organizational Excellence workshops that we hold at National—the two- or three-day quarterly off-sites—can work just as well for almost any size company, a division or work group, or smaller organization.

The trick lies in scaling the operation down to the need. Think in terms of a one-day workshop, or even, if necessary, a regularly scheduled morning or afternoon. A resort is great if it suits the style and budget of the company, but you can do as well in a meeting room of a nearby restaurant or hotel, or even a conference room or classroom without telephones in your own facility. Get everyone eager for a learning, growing experience by transforming the perception so people see the sessions as a *reward*.

Make the atmosphere as informal as you can—if off campus or over a weekend, everyone should be dressed casually. When possible, provide opportunities for volleyball games (a favorite of mine) or other activities that build a sense of working together toward a common goal.

Mealtimes should be structured in a way that encourages people to talk about the larger external issues and topics, not the daily business routine—preferably getting to know each other in ways that they may not have found the opportunity for in the pressure of the everyday scurry.

MAKING OE WORK

Here are a few of my personal tips on making OE work:

- Look for opportunities to bring in outside experts, advisers, and lecturers. The account manager who wants to increase penetration can set up a one-day session at which an invited lecturer talks on the leading ideas and new thoughts on the subject; this could be a professor from a local university or an author who has written about it. While most consultants will of course charge for an appearance of this kind, other experts like the professor or the author will often gladly appear without a fee.

  Another excellent possibility is to invite somebody from another company. At National, we've had speakers from companies like Hewlett-Packard. Because we're a major HP supplier, it's to their advantage if we get better, and so worth their effort... and we do the same thing with companies that sell to us.

- For these OE sessions or off-sites, I generally pose a single question for the group to tackle—one that focuses on a specific goal. A manager's question might be, "How do we get account penetration from 10 to 20 percent?" I usually don't announce the question in advance, because I find spontaneous discussion to be far more productive than the "prepared statements" people may feel compelled to have in their back pockets if they think you're expecting them to be cogent on the topic.

  Once you've introduced the question, you need to bite your tongue and refrain from telling them how you think they should get there.

- Let your group set their own priorities to the maximum extent. Your role is to guide them toward their own solutions, intervening only where they drift, bog down, or come to impractical solutions. You're the cowboy nudging steers, so they don't stray from the route.

- When they decide, let them implement. They will see you as empowering leader instead of an autocrat.
- The tendency is to drift from the predetermined goal—losing focus, clarity, or intensity. Your chief task is to make sure they stay aligned.
- For the middle manager, my advice is: In the beginning, give your subordinates free rein to their energy. Granting them a little too much freedom is better, and in the long run far more fruitful, than granting too little.
- Finally, an essential here as elsewhere in management: Every session must end with a list of action items. Enough said.

WRAP UP

Transformation requires more than giving speeches about "Win one for the Gipper." Organizational Excellence provides a process for transformation—a method for achieving quantum change without the danger of an unacceptable level of risk.

It's not possible to over-emphasize the importance of using a self-consistent framework such as the 7Ss. In addition to being a tool for organizing your thinking, a framework provides an essential checklist for change.

One of the most important lessons that a framework such as the 7Ss forces on you is the knowledge that you cannot fix one thing at a time—you must fix the *system*, dealing with all elements simultaneously.

You want your people to be innovative in a management sense. Encourage them to come with ideas they didn't get from you, and when they do, celebrate it. This is what you're hoping to nurture.

The ability to deal effectively with the soft Ss is what's going to make you great. In football they say, "Offense wins games, defense wins championships." The soft Ss are your defense.

If you can get people to the point where they are making the same decisions you would, they're going to be making 10,000 good decisions a day instead of ten. That's transformation... that's the multi-dimensional manager... and that's how you win.

END NOTES

1. Crosby, Philip B., Quality is Free, McGraw-Hill, 1979.

2. Quotations and figure from In Search of Excellence: Lessons from America's Best-Run Companies, Peters, Thomas J. and Waterman, Robert H., Jr. Copyright © 1982 by Thomas J. Peters and Robert H. Waterman, Jr. Reprinted by permission of HarperCollins Publishers, Inc.

3. "Systems" was originally included by the McKinsey team with what we now call the soft Ss, but is today more appropriately grouped as here, with the hard.

13

Operational Excellence—The #1 Strategic Weapon

"Operational Excellence is all about the ability to implement."

In five years Gil Amelio had led Rockwell's Semiconductor Products Division (SPD) from annual losses as high as $20 million to become the world's leading supplier of the electronic brains for the fax machine. At that point Don Beall, then COO of Rockwell, invited Gil to take over a much larger segment of Rockwell business that was similarly troubled, and try to perform the same kind of magic.

The move took Gil from a Rockwell division then doing $300 million in annual sales, to become president of an $850 million operation based outside Dallas. Called Rockwell Communications Systems, it included, in addition to SPD in Southern California, two divisions new to Gil: Switching Systems, in Chicago, and Network Transmission Systems, co-located with RCS headquarters in Texas. Both of the divisions being added to Gil's authority were in trouble, and once again he was being challenged to pull a rabbit out of the hat.

Each organization has its own fingerprint, its own arcs of strength and whirls of weakness. Five years earlier, Gil had quickly discovered that SPD was grimly focused on a dying market. What would the problems turn out to be in the two divisions he was just taking over?

Don Beall was clearly wondering the same thing. Beall "loves getting up to his elbows in strategy," and after Gil had been in his new position just a few months, his new boss arranged for him to appear before Rockwell's top management to explain the strategy he had evolved for the rescue operation.

At the presentation, Gil put up a pair of slides; the first one said—

The #1 strategic weapon is...

And the second one finished the sentence—

Operational Excellence

Gil made sure they quickly understood what he meant by the term and the importance he placed on it, and then went on to fill in the details—the networks division was in the right business, with good products and adequate marketing... but they had been doing "a rotten job of implementing." From taking the orders, to producing reliable goods, to shipping on time—in virtually every aspect of operations, they were failing, and failing badly.

As for the switching division, their problem was related to old products: in the business of making massive switches for telephone company offices, the division had not introduced a new product in ten years.

Gil then launched into a description of what was, in essence, his first full-blown Operational Excellence program—detailing the steps he had already begun for getting support at all levels toward solving the problems.

At the time the "strategic weapon" line was a handy device for getting the attention of his listeners. It's since become a personal buzz-phrase for Gil, and accepted wisdom at each operation Gil has run, and, as well, a motivator in getting people to focus on this central issue.

One of the challenges that faces new managers taking over an existing workgroup or organization is how to get up to speed and assert authority, while not interfering with what's working successfully. "Muscle-flexing" courts disaster.

Like most successful leaders, Gil is a master at asking non-threatening questions. He advises new managers to "keep from getting focused too soon," and it's typical Amelio style to dig for business information while at the same time learning to work with his key people by seeking to understand how they make decisions. He says—

Conducting a business review allows you to learn about the business and about the people at the same time; I recommend calling for a review of the most recent quarter. By listening and questioning, you get the strong clues you need about where the major disconnects are—as well as learning a good deal about the strengths and weaknesses of the people.

OPERATIONAL EXCELLENCE—WHAT IS IT?

The previous chapter described *Organizational* Excellence as our way of aligning effort and providing focus to make transformation happen.

Operational Excellence deals with the *how* of the business—all the programs and action involved with producing the goods or providing the services. It speaks to a combination of productivity and craftsmanship.

The formal definition we use at National is—

Operational Excellence

A high level of organizational effectiveness in delivering value to customers worldwide, resulting in superior gross profit performance.

Clearly a great variety of elements are wrapped up in this definition—from cycle time, to service, to nonstop quality. We aim to offer a disciplined but thoughtful mindset that involves more *how* than *what.* One of the National executives said that our underlying attitude for Operational Excellence is, "Good enough, isn't."

"THE NUMBER ONE STRATEGIC WEAPON"

The reason I talk about Operational Excellence being the number one strategic weapon is because of the notion—which I would think was obvious if it weren't so often overlooked—that having a great strategy is worthless if you can implement it. Operational Excellence is about the concerted efforts for implementing the programs, processes, and visions that will turn your strategy into reality.

Think for a moment about the effort to obtain a patent: you need to convince the patent office of two things—that your invention is not obvious; and that you are showing either a reduction to practice (an authentic way to make the idea work), or a way that someone else could pursue to reduce it to practice.

Likewise in business, you don't have a valid and useful strategic plan until you have thought through and can be cogent about how to put it into practice. So Operational Excellence is the "reduction to practice" of strategic planning—the descriptions of how you will achieve the strategic goals.

Operational Excellence and strategic planning are the yin-and-yang of success in business. Strategic planning defines the direction in which you're heading and the battles you're going to fight along the way. To put it in more practical terms, strategic planning is about deciding how you're going to focus and allocate the resources of the company. Many of the key decisions needed to achieve Operational Excellence emanate from the strategic plan.

You can't accomplish Operational Excellence merely by encouraging people to work longer or harder. It's a common misunderstanding to think, "We need to change, we need to improve, everybody needs to try harder." Although the cliche says there's always room for improvement, trying harder without a process to support the goal simply leads to frustration. Like a New Year's Eve resolution to lose weight—until a diet plan or weight-loss program has been chosen, the resolution sits there, going nowhere. (A problem many of us can attest to—me included!)

At National, part of the required process is learning how to be excellent—discovering what's required to create a learning organization, which involves using the tools discussed earlier—training, feedback loops, and so on—so employees achieve responsive change.

It takes longer to change perception and style than just *telling* people what to do, yet without evolved perceptions, the behaviors don't hold and problem solving is less consistent. Strong leaders who are genuinely working for transformation know that their people need to sense the theories and thinking behind the instructions. Transformation leaders, in my view, must be encouraged to develop a realistic and practical quantity of patience. The fast action type, the "do like I tell ya" bully style of leadership, isn't part of the transformation process. Leaders or managers who like this style often are very bright and very eager people who, when they learn some patience, can often become transformation managers.

Talking strategy is admirable and highly desired, but this must exist in tandem with the capability to implement, and as I've indicated, the high-level process you use for implementing Operational Excellence is *strategic planning*.

THE BIRTH OF STRATEGIC PLANNING AT NATIONAL

Many people—including a good many business school professors and, as a result, their students—assume that strategic planning is a natural, routine part of every ongoing business. They're wrong.

National did not have a strategic business plan when I arrived, and that's casting no aspersions on the company. The failure to do meaningful strategic business planning is a widespread affliction. When I asked the top managers and product line managers to do strategic plans, they didn't have any idea where to begin.

At that point I sent out a call for help to Michael Townsend, who had been my vice president for strategic planning at Rockwell. Mike's company, Decision Analysis Corp., is now based in Oregon.

Townsend starting working with National's people from the very first Executive Off-site, where he addressed the basics of strategic planning: why companies do it, how it's done, and how it can be kept from becoming a degrading and pointless waste of time. It was National's first taste of the many strategic planning tasks and process that lay ahead.

Still working with us on a continuing basis, Townsend has guided the design and evolution of our strategic planning process—writing the process description (which became famous around the company as the "blue book"—excerpts are presented in Appendix E) and a number of related tools and examples. Over several years, dozens of National managers have gradually become skilled and comfortable at devising the long-range plans needed for profitability and growth.

THE STRATEGIC PLANNING PROCESS

Perhaps the most frequently asked question put to me by other CEOs is, "How do you do strategic planning?" On the threshold of a planning cycle, many experienced people start having nightmares.

At National, we follow Mike Townsend's "Prolog/Dialog/Postlog discipline, in which the company kicks off the process by making a clear statement (the Prolog) of the very few real constraints that we want individual business managers to pay attention to—the amount of capital or R&D investment we want to make in a given business, for example, or the level of sales we expect.

Next comes a series of interactive discussions (the Dialog) between the business managers and the managers they report to—typically division general managers and group presidents. In this process our business teams develop a coherent vision of the future, a set of objectives, and the strategic alternatives from which a specific best-choice strategy will be chosen. The business strategies themselves are articulated in terms of what Townsend calls "strategic action statements"; these are specific things the business is going to do, and where possible, we expect a specific time frame and a measurable outcome.

After "rolling up" all of the individual plans into a corporate plan, final decisions are communicated back to the business in the last planning phase, the Postlog.

Individual business strategies and forecasted outcomes are rolled up in the usual way into division and group plans, eventually culminating in the overall company plan. Naturally, in the course of this rollup, various "adjustments" have to be made—when we feel a given business isn't being aggressive enough, for instance, or where we feel a particular strategy involves more risk than we want to take in a given market arena. In any case, these inevitable adjustments are communicated back to the businesses (the Postlog) so that individual business managers will have one more chance to plead their case, and where necessary, adjust their plans.

In another instance of how we practice empowerment, we try to push the strategic planning process down to the lowest practical organizational level, which we've dubbed the "strategic business element"—often an individual product line or similar business unit. While this makes the rollup a bit more cumbersome and time-consuming, the resulting plan ownership throughout the company is more than worth it. We also ask our sales and service arm, and various technology-development organizations, to participate by making their own plans, which are expected to dovetail with those made by the business organizations.

While this effort is messy and tough to coordinate, and often leaves some loose ends and ambiguity, in the final analysis getting everyone involved and thinking in the same direction is infinitely more important than the cosmetics of the plan itself. In the words of General Dwight Eisenhower, "Planning is everything. The plan is nothing."

The strategic business elements and the sales and technology groups are the building blocks of a company which, when assembled, should form a handsome edifice. Unfortunately you'll discover, as we do each year, that what they form is more like a Swiss cheese, full of gaps and holes. But these holes help us choose where to focus executive attention, and this circles back to the question of operational planning and excellence: what do we have to repair, install, upgrade, renew, acquire, in order to shape the destiny of the company consistent with our overall objectives.

Here are a few of my tips to help with the strategic planning process—

- As the manager, it's up to you to conduct strategic planning meetings in such a way that all ideas and plans presented will, indeed, support the objectives of the company. (It helps to have the company goals, objectives, and Vision statements clearly displayed in the meeting room. A "Pledge of Allegiance to the Vision" would be a silly reality, but is symbolically what you're after.)

- Be sure the cut-off date for input into the strategic planning process is well known, so people don't keep trying to input ideas when further additions just complicate the issue. Even the best idea submitted after a certain point in time becomes a burden. Get your people to understand that an idea is either a benefit or a burden depending on when and how it's submitted. (You might ask, if it's *planning* and not the *plan* that counts, why is a cutoff date important? The answer is that real strategic planning is—or at least should be—something that goes on all the time. The plan document is the once-a-year setting down in words of your current stage of thought. So a cut-off date is a practicality that says, "as of this date, here was the status of our thinking." To your people, you're saying, "If you miss the train this time, don't worry, there will be another coming along shortly; in the meanwhile, keep thinking and working.")

- Sometimes it's necessary to help people be certain they're in the right forest before they start talking about the trees. In a recent round of strategic business planning at National, I announced that people were to come in with *no* charts, slides, or overhead-projector foils. With a set of visuals to lean on, the tendency is to include a bullet point for every item of the program, and then talk to every bullet. When the presentation is entirely oral, most presenters realize they need to keep to the high-level points that are the real concern. And it also conveys the message to your people that the quality of the strategic thinking is what's important, not the fanciness of the slides.
- In any planning process, always make sure the "road map" goes far enough out into the future. With product plans, I like people to look outward for *three* generations. Before I approve a new product development, I want to know what the two subsequent generations will be. By looking further ahead I have found that the *first* product in the family tree is always improved.

HOW MANAGERS CAN HELP ACHIEVE OPERATIONAL EXCELLENCE

In business, in personal life—in *all* decision making—we're always in danger from the overlooked items that end up biting us in the backside. If there are factors at play that you're not even aware of, the odds are against your ever being successful. At National, we refer to these items, using a term I mentioned earlier, as "things that fall into the white space"—meaning factors that weren't addressed as the responsibility of any particular person or group, and so never got noticed or attended to.

Clearly the Catch 22 is, How do you identify in advance the items that are likely to be overlooked? Being an old programmer, one method I personally favor is to draw a flow diagram even though I recognize this leans toward a conceptual view rather than a literal one, which is not always an advantage.

A process we use more widely at National involves something we've come to call the "Is Map." You create an Is Map by drawing a flow chart that provides answers to the question, "How are we currently doing this operation?" And then you create a companion statement, which we call the "Should Map," where you provide answers to the question, "How could we be doing this better?"

The process is revealing. Virtually every time we go through these steps, we find ourselves scratching our heads and wondering, "Why on earth did we ever end up doing things *this* way!"

You need to be constantly on guard in every part of your work, regularly challenging yourself—"Am I overlooking something?" "Have conditions changed?" "Are things creeping into my process flow that are deteriorating the process?" "How recently have we checked to find out about new techniques or knowledge that we haven't yet begun taking advantage of?"

As one real-life example, I recently noticed that National was repeating a bizarre condition in which we had a rising number of failures in the ability to fill orders, and rising inventories, at the same time. That sounds contradictory, but it's what happens when you build the wrong products. This is the worst of both worlds—too much inventory, but unable to fill customer orders.

Expressed in this straight-forward way, the situation seems obvious, but in fact it happened two or three times before I sat down and asked myself, "What are we missing here?" To simplify a complex story, the answer was that we were overlooking two very important aspects of inventory management—we did not have a good model to drive the operation, and we did not have a good demand forecast system to give the factories a prediction of what we're probably going to want them to produce over the next twelve months.

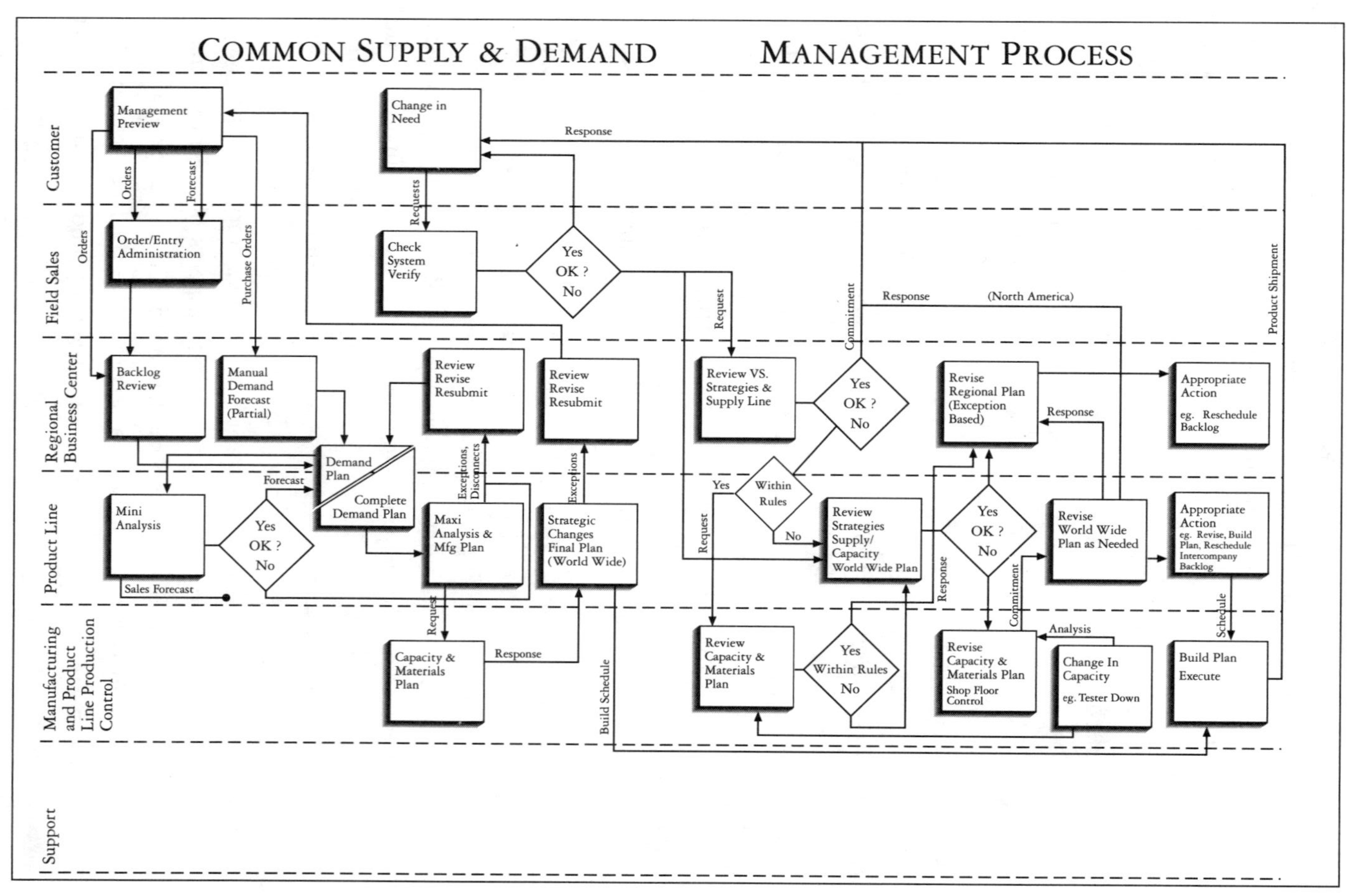

Sample "Is Map" flow chart.

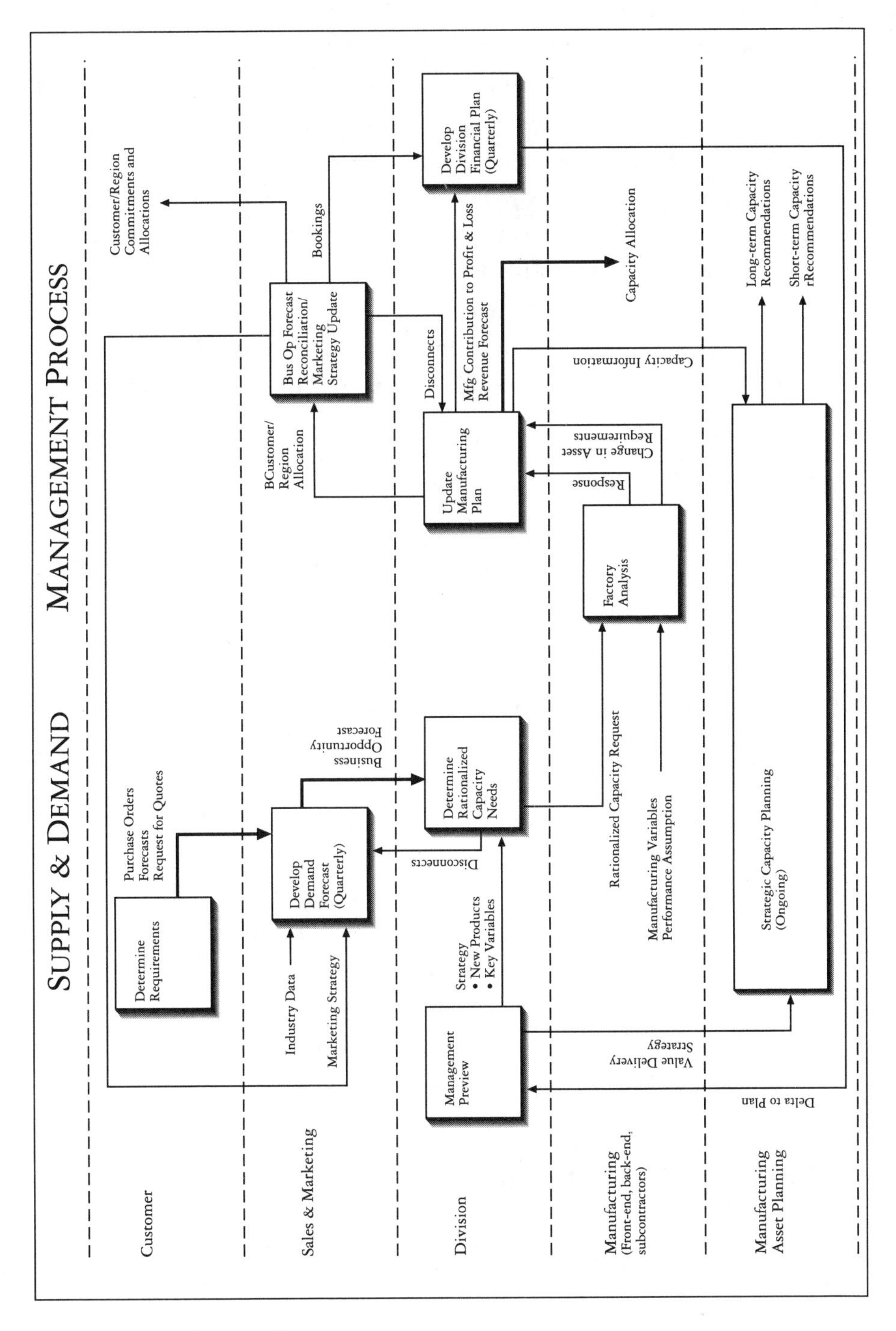

Sample "Should Map" flow chart.

This is another aspect of a behavior I talked about earlier. I recommend developing the attitude of being a natural skeptic rather than making rapid assumptions without getting enough information. High-powered managers and leaders are a valuable asset to a company, but they often shoot themselves in the foot by drawing a conclusion too quickly. It was said of President Teddy Roosevelt that he "thought in his hips." President Dwight Eisenhower, on the other hand, said that a fast "yes" or "no" is not always a display of leadership; there are times when the best course is to wait for better information before making a decision. ("No decision is sometimes the best decision" is, I believe, the way Eisenhower put it.)

MAKING BETTER USE OF MIS

Another arena where managers can take personal action to support Operational Excellence is in their use of material from the company's Management Information Systems (MIS).

A management information system that puts the right information onto the right desk at the right time is an extremely important part of making things happen. But an information system can also be a waste of effort unless managers are using its full potential.

Most of the MIS organizations I've had any dealings with are delighted to have users come to them. I've formed the impression that they sometimes feel undernoticed or at least underappreciated, and when someone shows up at the door with a specific "Here's what I need, can you do it for me?" they're only too happy to accommodate.

So go to MIS and tell them what information you want... but only after thinking your needs through very carefully (see "What you measure is what you get," in Chapter 17).

RECOGNITION AND REWARD

A final item I would include under this topic of "What can a manager do to implement Operational Excellence" is recognition and reward.

If you establish metrics and benchmarks (another topic covered in more detail in Chapter 17) on some aspect of your competitiveness—such as yield or customer returns—and then reward and recognize people when they do significantly better than the competition, this becomes another step that will help you achieve implementation.

Reward and recognize, and celebrate success.

HARD-EDGED MANUFACTURING MANAGERS

People at National have heard me say that "I love manufacturing managers who have hard edges"—referring not to bullying types, but to people used to seeing the world in clearcut terms. Certainly that's not what I want in every company position, but in manufacturing, I much prefer people who see the world in black and white, rather than in shades of gray.

Why is this good? Because excellence in manufacturing is attained in large measure by dealing with a myriad of small problems that are constantly nipping at your heels. Excellence comes from being very rigorous at attending to the details.

An example comes to mind from an incident that just recently occurred. At our wafer fabrication plant in Scotland, some operations were yielding well (producing a sufficiently high percentage of good chips), and others using the same process and materials were not. Why the difference?

A detailed analysis traced the problem back to the two "implanters"—machines that use an ion beam to "implant" electronic paths on the surface of a chip. The machines were identical units, from the same manufacturer. Further probing revealed that the angle of incidence of the ion beam relative to surface of the wafer was a mirror-image setting—45 degrees in one machine, 135 in the other.

More questions eventually turned up a good technical reason for the different settings—when the newer of the two machines was originally set up the same way as the previous one, it was breaking wafers. The change of implant angle solved the breakage problem and appeared in all other respects to be equivalent. But in fact the ions, instead of being absorbed at surface, were traveling much deeper into the wafer. Armed with this understanding, a way was found to reset the machine that avoided both the mechanical problem and the ion-penetration problem.

The matter was solved only as a result of thorough detective work, the equivalent of knocking on 100 doors to get a piece of information that solves a crime. Manufacturing involves an unending stream of similar challenges, which is why I believe that "hard edged" people—who don't take anything for granted—are necessary in this part of a business.

Personally, while I understand the technology, I don't get my greatest pleasure from digging into the minutia. But to some engineers, finding something like the implant angle is what they get up for in the morning. People like this provide the only way you can achieve Operational Excellence in the manufacturing part of your organization.

WRAP UP

While Organizational Excellence provides a way of aligning effort and providing focus to make transformation happen, *Operational* Excellence deals with the *how* of the business—the specific programs and actions you implement in order to produce the goods or provide the services. Look on Operational Excellence as your company's number one strategic weapon, because even a great strategy is worthless if you can't implement it.

The quest for Operational Excellence needs to begin with a well organized strategic planning process; strategic planning is about deciding how you're going to achieve Operational Excellence by focusing and allocating the resources of the company.

National's strategic planning is based on a three-phase process. In the Prolog, we set out the constraints we want individual business managers to pay attention to, such as the amount of capital that will be available to a particular business. The next stage, Dialog, takes place as a series of discussions between managers of the individual businesses and the managers they report to, aimed at developing a best-choice strategy for each business unit. Once the individual plans have been incorporated into a single, all-encompassing corporate plan, final decisions are communicated back to the individual businesses in the last planning phase, the Postlog.

Some questions to ask yourself during strategic planning—

- Do all of the plans we are proposing support the objectives of the company?
- Have I let my people know the cut-off date for input, so they can get their ideas submitted on time?
- Have I made sure my people are using a "road map" that looks far enough into the future? (For products, three generations is not too far ahead for planning purposes.)

One important way that an individual manager can contribute to Operational Excellence is by consistently checking his or her operations to make sure that nothing is "falling into the white space." At National, we use the "Is Map/Should Map" process as a way of helping people to organize their thinking.

Some aspects of transformation can be set aside once you have achieved an initial organizational equilibrium and created the previously missing programs for communicating, rewarding, and so on. But Operational Excellence is not one of these. Instead, it's something you always need to be working on, through every phase of transformation.

Once you have achieved a measure of Operational Excellence, you and your managers will have a greater degree of freedom—freedom in deciding issues like what markets you might best go after and what customers you might best be able to serve.

14

Teamwork at the Top, Teamwork Everywhere

"What managers manage is change. The rest is administrative."

In one form or another, all companies use teams. One team at National through the late 1980s, led by the president of National's International Business Group, Pat Brockett, was struggling with what Brockett calls the "difficult to impossible" task of developing strategic programs with large customers.

The problem was that these customers were savvy to National's condition. Brockett says, "You knew they were questioning our ability to finance the product development, our production capability, and our very survival. And obviously this had an impact on the morale and stability of our sales organization."

One particular incident sums up Brockett's sense of the changes at National.

> *We used to go at least once a year to IBM corporate procurement headquarters in Poughkeepsie, New York. Most of those interactions were fairly unpleasant because we were falling well short of IBM's expectations in quality, service, or both. So they were not particularly pleasant trips for us.*

After Gil Amelio took over, he and Brockett visited IBM together on what turned out to be an unfortunate introduction for the new CEO.

> *IBM put National on probation as a supplier for an ongoing quality problem caused by the ancient molding process we were using. Although it was never directly said, I knew the customer really felt we didn't have the cash to make the appropriate investments.*

Despite IBM's concerns, the National team addressed the difficulties, met IBM's requirements, and stayed on as a supplier. Two years later, Brockett was back in Poughkeepsie for another annual session. By then—

> *we had several quarters of sustained profitability behind us, our quality rating with them had improved, and we had made significant investment to upgrade the mold we used in processing their products.*

I arrived at the small old hotel we usually stayed at, was met by my very enthusiastic sales team and taken into the bar where there was a large roaring fire. We had been talking for about thirty minutes when my IBM sales manager, Gary Saunders, looked down at the end of the room, turned to me and quickly said, "The TI team is at the end of the bar." He waved to acknowledge the group.

The four of them apparently wanted to meet me; they came over and introduced themselves, and one of them said, "You guys have been doing a hell of a job at IBM. We were in there today and they held you up as an example of the supplier we should aspire to be."

Another said, "You at National have done a phenomenal job in the last couple of years turning this around. You must be feeling great."

I'll never forget the look of pride in Gary Saunders' eyes.

THE VALUE OF TEAMS

A team curriculum originally developed at National Semiconductor as part of a Total Quality program eventually fell victim to flagging enthusiasm and the company's shortage of resources. When the effort was eventually brought back to life, instead of limiting it to project and development teams, we resolved to view teams as instruments for transformation and change.

On a corporate level, teams offer a way to achieve greater flexibility and responsiveness. For the individual, they offer a healthy variation from the everyday routine, an opportunity to develop and gain recognition for new skills and talents, and a way for people to put their ideas and insights into action.

We've developed a list of benefits from using teams which falls slightly short of a David Letterman's "Ten Reasons Why..." list, but gives an indication of the reason we depend on them so heavily:

9. More people in the organization share leadership.
8. More experience and knowledge is brought to bear on complex problems.
7. More information goes directly to the point where action can be taken.
6. More decisions are effectively and quickly implemented because the implementors have been part of the decision making process.
5. More employees view themselves as business partners striving for the same goal.
4. More innovation and risk taking is possible because there is a higher level of trust among all members of the organization.
3. More people are listening to customers and have authority to solve customer problems.
2. More people focus on learning from mistakes rather than unproductive competition between groups.
1. More global thinking and long range planning takes place because, as teams make more day-to-day decisions, front-line leaders have more time for the higher-level tasks.

The bottom line is that we now use teams much more widely, for more purposes, and grant them greater authority for action.

TEAMS, NOT "TEAMWORK"

Teams represent a practical and often very effective tool; what passes in the name of "teamwork" suggests a company that is looking for all lightness, harmony, and love. The cliche about "being part of the team" is understood to mean being nice, everybody getting along well.

As this chapter makes clear, what can be achieved with teams has little to do with the thin soup of pretend fellowship, which is what too often passes itself off as "teamwork."

THE LIFE CYCLE OF A TEAM

One of the great dangers in relying on a team approach is that teams often tend to be like programs of the federal government—once created, they may remain in existence long after they've ceased to have a benefit that's justified by the amount of effort going in. There are companies where it's next to impossible to get anyone on the phone, because the entire staff seems to move from one team meeting to another—stopping by their desks only long enough to read their e-mail.

It seems to be a law of nature that teams acquire a life of their own, and in the long term become destructive to the progress of the organization.

To avoid this, I insist that, with few exceptions, a new team must be given a self-destruct timetable—something on the order of 90 to 120 days. That should be long enough for the people to achieve something useful, and move on. (Don't we all wish a lot of government programs would be handled this way!)

GETTING A TEAM STARTED

Out of our experiences at National, we've been able to formulate some useful guidelines for using teams successfully.

The decision to create a team to address a particular issue should be made by the management of a division or product line, when one or more of these conditions occur—

- the problem or issue can't be readily resolved by individuals acting on their own
- a balanced solution to a complex problem needs to be found quickly

- the problem involves specialized, cross-disciplinary expertise.

A particularly powerful tool we've developed at National lies in the use of what we call "sponsors." Once a management group has decided to form a team to address a particular issue, a member of that management group is nominated to become the team's sponsor; he or she assumes responsibility for the success of the effort.

The sponsor charters the team, selects a team leader, gets them resources, and defines their authority. The sponsor and team leader together decide who's most appropriate to be on the team.

You can use teams to address problems ranging from the corporate, global level, to specific matters on the shop floor. And the leaders need not be managers: at National, of the hundreds of such groups operating at any one time in the company, a great many—indeed, probably most—are being run by a team leader who isn't a manager, and who, although trained in a team leader course, may have limited experience in the role. These people gain practical experience in leadership while receiving support and guidance from the sponsor, who throughout tries to clear away the obstacles in their path so that they have the best possible chance of being successful.

TRAINING FOR TEAMING

One reason teams are not more successful in some companies is the unspoken assumption that, for the average manager or worker, team participation fits the description in the line from *Annie Get Your Gun*—it requires only "doin' what comes naturally."

Under Kevin Wheeler and Bill Schlecht, National has developed training programs for both leaders and participants. Team leaders attend a five-day course offered at National Semiconductor sites around the world. In addition, we also offer more than a dozen specialty modules based on the Zenger-Miller material.

Most National employees have the opportunity to serve as a team member, and probably 90 to 95 percent of our people—from production-line workers to senior managers—have been through one of our team-participant training programs.

THE FOUR AGES IN THE LIFE OF A TEAM

An effective team—especially one that is limited to a three- or four-month lifetime—should travel through the four phases of what's become known as the Tuchman Model: "forming, storming, norming, and performing."

Forming is the process of getting people together, making linkages, building relationships, and finding common ground. The Storming stage may, for the uninitiated, come as a surprise—it is a period during which the participants are *expected* to disagree. It's encouraged, because we want people to express their views without reservation, to disagree, to dispute one another's viewpoints, and to try to persuade.

In the Norming stage, the group comes to an agreement on a plan or course of action. And the final step, Performing, carries out the new plan.

CAUTION: TEAMING CAN BE ADDICTIVE

Once you have an effective team process up and running, your people trained, and the 90-120 day time limit firmly entrenched, you may have one more hurdle to worry about: team participation can become addictive.

Some people get turned on by the success of teams in solving problems, and in the satisfaction of working together to find solutions. You turn around one day, and the organization is launching a team for *everything.* When you launch teams inappropriately, you're in danger of landing back in the bureaucratic morass, because you can't do anything until fifteen people have voted in favor of it. So the first question that one must ask becomes, "Is this a problem that is better addressed with a team, or with individual action?"

COUNCILS

While teams at National are designed to have a short life span, we also have another, corresponding type of program that is long-lasting: the various "councils." The major distinction is that unlike teams, the councils are largely charged with setting policy. And they also provide another highly valuable, ongoing level of feedback loops. In the process, they create a harmony all across the organization.

It used to be that many of our general managers were virtually strangers to one another; that's not surprising—they generally had only occasional need for contact, some were at far-flung locations, and they came together only on special occasions for company business in a large group setting.

The Executive Off-sites have helped create a greater sense of community and unity of purpose. But beyond that, these same business managers also gather periodically in a group that we call the "GM Council." They share ideas, and then agree on a project they want to promote for the entire company. One of their projects was to modify the various Activity-Based Costing programs around the company into a uniform version that would be consistent in all divisions and departments. On another occasion, the GM Council set up a project dedicated to improving quality across the entire company by a factor of two.

We've also created a Quality Council, which similarly provides a forum for discussing issues and finding solutions to quality problems that are widespread, rather than tackling them as a dozen separate challenges in a dozen parts of the company. This is especially valuable at a time when we are under pressure from customers to meet the standards of the particular quality program they subscribe to; Ford wants our automotive division to adhere to the quality standards of ISO-9000, and so on. The Quality Council addresses questions such as how to meet this type of customer demand, and yet do so from the base of a consistent, company-wide quality standard within National.

Our Manufacturing Council (later called the Value Delivery Council), made up of the people who run our major manufacturing activities, deals with common issues such as training, scrap, and improving worker quality.

And we've also created a council designed to enhance the Leading Change process. The idea behind the LC Council was to provide a way for continuing to influence "graduates" of Leading Change.

Each class of middle managers, at the end of their LC training, was asked to select one class member who would become their representative to the Leading Change Council. The Council now has a membership limited to twenty-five people; based on input from their own classmates, the members gather to discuss and arrive at recommendations for improving the Leading Change process and training. They then convey these recommendations to the senior executive staff.

In this way the Leading Change Council has become an ongoing form of empowering the middle managers, and a continuing factor in transformation.

Within your own company, think of creating Councils where conflicting policies or practices in different parts of the company are impeding the progress of transformation. Also look for opportunities to create Councils that can enhance learning by providing a feedback loop, channeling to top management ideas for enhancing and speeding up the transformation.

CONSULTANTS

If its true, as I believe, that what managers manage is change, then the rest is administrative. If tomorrow were exactly like today, we wouldn't need management—we'd be stuck in a feudal society, where the son did the same job as his father, and no one had to tell him how to do it.

The need for management comes about when the organization must create new products and market them. That's when management energy most needs to be devoted to managing change.

Even in a relatively small organization, creating change is like turning a battleship. In order to change rapidly or radically, you need to put more energy into the system. The pressure of transformation brings with it the challenge of seeing that the business of the organization continues to be achieved, while at the same time everyone is expected to learn new ideas and attitudes, and develop new ways of functioning.

When you want to increase the rate of change, you need more management energy injected into the system. But no company today can afford to have an army of managers waiting in the wings to be assigned.

One powerful way of enabling the corporation to create rapid change is to enroll the aid of consultants. At times of the greatest need for change, I enhance the capabilities of the corporate team by calling on the talents of a carefully selected group of consultants—people screened by our experience with them or information about their background, and chosen for their specific talent to help augment the existing management teams. This allows us to maximize the energy at a given point of focus. And after the organization is successfully on a new tack, we're then readily able to reduce the cost to competitive levels, simply by ending the consultant's assignment until we next have a need for those particular abilities.

In selecting consultants, I stay alert for people skills as well as content capabilities—for a style that will support my goal of continually growing and transforming people. I seek and measure as successful those consultants who work shoulder to shoulder with my people—imparting knowledge, providing guidance, and, by supporting them, helping the company people to succeed. The best consultants have the spirit of the teacher imbedded in their style and help people learn in the way we all know is best: by doing and discovering for themselves. There's a clear difference between the contractor brought in to do a specific task, and the consultant who works with your company over an extended period. It's important not to confuse the two by measuring their performance against the same set of criteria.

(I confess I've also heard it said about me that I prefer consultants who are willing to remain in the shadows and let others take the credit!)

CONSULTANTS FOR THE SMALLER ORGANIZATION

As suggested earlier, the roster of consultants available to you ranges from local college professors, to individuals (who these days may be someone who held a good job somewhere until he or she was "re-engineered" or "down-sized" into the world of the not-regularly-employed), to one of the organizations large or small that are in the business of helping businesses solve problems.

Advice is, of course, easy to get; the challenge is, how do you make certain the people you choose to advise you are truly capable of providing valid, useful solutions?

I offer two answers: Ask for recommendations from others whose opinions you respect. And (especially for younger managers) start working with consultants before you have a dire need for them; build relationships over the years so you have a small pool of people you trust and know you can depend on when an urgent need arises.

CONSULTANTS AT NATIONAL

Not having had anyone offer me this advice early in my career, I count it my good fortune to have picked up a small number of people along the way, on whom I've been able to depend in years past, and in helping with the transformation at National.

Two of them have already figured into this story—Professor Bob Miles of the Emory Business School, and Michael Townsend, of Decision Analysis Corp.

A third is McKinsey & Company, who have been very helpful on problems involving shifts in strategy. For example, National had been trying to implement a strategy for acquiring major customers in Japan, but getting nowhere. McKinsey's Bob Attiyeh, then a long-time team leader from McKinsey, thought through the problem with our people, led us to develop a more effective approach, and helped us through the first few months of implementing it.

Some people call on consultants only when they face a sweeping mission—"My factory is broke—help me fix it." Or, "We're not in the right businesses, tell us what business we should be in."

That's an abdication of the managers' responsibility. Consultants do the most good when they work shoulder to shoulder with strong managers. Use them as advisors—making decisions *with* you, not for you.

WRAP UP

Any organization that is not using teams extensively and effectively is flying with one engine on idle. Teams offer a way to achieve greater flexibility and responsiveness for the organization, while taking greater advantage of the talents of the employees.

At National, people frequently serve on three or four teams a year. In addition to other benefits, the opportunity to work with people from various parts of the company—sometimes from National sites in other countries—provides another way we establish linkages throughout the organization, a valuable benefit I've talked about earlier.

Team participation also teaches people the art of amplifying their own talent by using the talents of others—a highly valuable capability to pass along to the younger and the lower-ranking people in your organization.

Those chosen to serve as team leaders gain a chance to exercise authority; for people who are not managers, it's an opportunity to learn in small doses some of the skills needed for advancement. At the same time, company managers get a good opportunity to spot promising talent.

For creating teams within your own company:

- provide training in being a team leader and in being an effective team member
- use a team when the problem or issue cannot be readily resolved through action by individuals, or when a solution to a complex problem needs to be found quickly, or when the situation involves specialized, cross-disciplinary expertise
- whenever practical, don't select team leaders from the manager pool, but instead use the opportunity to provide leadership experience to supervisors or others
- assign a Sponsor to each team—a member of the organization's management team who will provide support and guidance to the team leader
- establish a specific timetable under which the team ceases to exist after a certain date, preferably allowing a lifetime of about 90 to 120 days.

A team should be created when one or more of these conditions occur—

- the problem or issue cannot be readily resolved by individuals acting on their own

- a balanced solution to a complex problem needs to be found quickly
- the problem involves specialized, cross-disciplinary expertise.

The Sponsor provides guidance and the benefit of his or her experience in moving obstacles out of the way.

Especially in periods when the company is going through transformation or making other rapid changes, the effectiveness of teams and of the organization as a whole can be significantly enhanced through the use of consultants. In selecting consultants, be wary of the type who come in, look around, tell your people what they are supposedly doing wrong, and then disappear. Instead, look for consultants whose style is to work shoulder to shoulder with your people—imparting knowledge, providing guidance, and, by supporting them, helping the company people to make the decisions themselves, in the process gaining new skills. When budgets do not allow bringing in "A" List consulting firms, look around at the local universities and among individuals recommended by your peers.

SECTION 4

Time, Metrics, and Money

15

"Speed is Life"—Using Time As a Competitive Weapon

"How do you use time as a competitive weapon? The answer is, you use it to increase your rate of learning."

An important challenge in transformation is making the company more adept at responding quickly to existing conditions and changing circumstances. But for a company like National Semiconductor, there is another aspect to this: a Fortune 500 company doing $2 billion annual business, National was barely known outside the semiconductor industry.

Part of a company's reputation lies in the numbers that give the measure of its achievement, but part lies also in the perception that the company creates about itself. Could we create a fresh perception of National—a perception of a nimble company capable of acting swiftly?

The challenge was presented to Anne Wagner, the company's vice president of marketing communications—

> *The problem was to create a communications strategy which would match National's business strategy, reflecting the new personality and positioning.*
>
> *While an outside agency worked on a new logo, the toughest part for my group internally was to define the new positioning with a short*

> *catch phrase that would accurately sum up the company, help ignite the work force, and tell everyone else how we're different. They spent a lot of energy trying to stamp out the right words. When we hit upon 'moving and shaping information,' it seemed to have the right combination of dynamic movement and solid business.*
>
> *Not long after, we put on a presentation to some market analysts and spent the morning trooping one product group after another past them. The analysts were thoroughly confused. 'Who is National?' they asked. 'What differentiates you? What are you all about?'*
>
> *Desperate, I grasped for 'moving and shaping' and informally talked through it in about five sentences. One analyst said, 'That makes more sense than everything we heard all morning.' I knew at the moment that we wouldn't have much trouble convincing management we had found the right words.*
>
> *After the external roll-out, the CEO of one of our competitors was making a public presentation to a group of analysts. I heard later that one of them asked him if his company was in moving and shaping. No, they weren't, but...*
>
> *Yes—our message was working!*

Change at National was being understood, accepted, and even embraced. Soon not just the employees, customers, and vendors, but the outside world as well, would realize that transformation was actually taking place. But it was not yet taking place fast enough.

Gil believes that time is a weapon you use against the competition, or the competition uses against you.

ANOTHER USE FOR FEEDBACK LOOPS: INCREASING SPEED

It's a familiar cry today that we need to shorten the development cycle, shorten time to market, and speed up the entire way we do business.

Many people make the mistake of thinking that shortening the cycle time simply means doing everything faster. The essential reason to shorten cycle time is not just because your company will be more responsive to customers, or because the amount of work per unit will be lowered, but because all experiential learning will be accomplished more rapidly.

It's worth repeating the point made in an earlier chapter that if your cycle time is short enough so that learning is happening faster in your company than at the competition, your organization will eventually pass the competition and win the business with better products and higher customer-satisfaction levels. Even if it takes several years for this to happen, the effort is well justified. This goal needs to be an intrinsic and carefully articulated part of the transformation process.

You'll recall the illustrative story (Chapter 10) about using a feedback loop to correct a problem with chips for NEC in Japan, so they would not continue getting defective material each month.

Feedback loops make you more competitive and more able to satisfy your customers because they improve your success at learning. At the same time, they also improve your ***speed*** of learning—which translates into speed of responding to customer requests, complaints, and expectations. How fast you can complete one feedback cycle and start the next is a measure of how fast you can satisfy the customer.

This is one of the reasons why the Japanese outran the U.S. semiconductor industry in the late 1970s, even though U.S. companies had created the market and dominated it for years. We were learning very slowly, while the Japanese companies had recognized the need to learn rapidly and had successfully developed effective techniques for achieving this. (More recently, in contrast, the Japanese seem to have slowed their rate of learning while the U.S. companies were accelerating theirs. Seen in this light, it is, perhaps, not surprising that the U.S. companies regained the lead in semiconductors in the early 90s.)

The traditional wisdom is that the first company to the marketplace has the advantage and will become the big winner. Not necessarily true. Certainly not true in the market climate of today; and even in the past, it was only true part of the time.

Consider IBM (which despite their well-reported troubles had 1994 revenues of $64 billion). Through the years, IBM earned a reputation for showing little interest in being first with new technology. They were successful anyway, and I believe it was because, once they committed to a technology, they put to work a set of methods they had developed internally for speeding through the learning curve. In recent history they've lost this ability and may have even lost sight of the importance of rapid learning. Some might theorize that by growing so large they lost the skills of learning fast enough to overtake the competition. I don't agree with that theory; more likely the leadership at IBM lost sight of the basic need for speed.

TIME AS A COMPETITIVE WEAPON

So—how do you use time as a competitive weapon? By now you know my answer: you use it to increase your rate of learning.

In a very real and practical sense, this is the only thing that matters. You can develop better products, you can be more responsive, you can have a corporate Vision that's more clear and better understood by your work force... but none of this adds up to much if you can't move fast enough to be competitive.

Winston Churchill once said that courage is the understructure of all virtue; in a very real sense, courage is the only quality that matters because without it truth, honesty, and loyalty are impossible to achieve. In a corresponding sense, I consider speed, especially speed of learning, to be the understructure of all individual and corporate strengths.

USING TIME TO PRIORITIZE

But how do you actually put "speed" to work?

Here's one example: Pick a process that you're responsible for, which you suspect of having a considerably longer cycle time than it should. The question then is, how do you go about systematically compressing this cycle time?

The answer is—In small steps… doing a series of individual things that will each force shorter cycle time… starting at the ***end*** of the process.

It's easier to picture this in a factory setting where your goal is to shorten the production time of a particular product. Start thinking about it, and you probably know two or three things offhand that will need to be done.

Begin by concentrating some extra management attention and energy at the end of the line, aimed at accelerating the output. When the velocity has picked up at this point, the next step is to concentrate on finding what it will take to speed up the process at a point a bit earlier in the line.

Continue to ripple backward in this way, pushing along and making improvements. Sooner or later a problem will be revealed—maybe congestion at one point in the process, or a piece of equipment that's too slow. Any further improvement in the cycle time of the process runs into the constraint of this particular operation.

The first time a problem emerges, make this the highest priority, focusing attention on it until you or the appropriate person has found a solution and resolved the problem—by rearranging work assignments, adding another worker, providing more training, upgrading a machine, buying new equipment, or whatever. Once this hurdle has been surmounted, continue working toward the front of the production line. When the next problem emerges, this new item becomes top priority until it's solved.

PRIORITIZING AND THE THEORY OF CONSTRAINTS

Anyone familiar with the "The Goal" (as set out in the book of the same name, by Eli Goldratt) will recognize that this theory says you should search through the ***entire line*** to find the worst constraint, and fix that first. In contrast, my approach says you should start at the end of the line and work forward. Despite appearances, these two are in fact not contradictory.

Almost any production flow, except for the most trivial, is really a sequence of work clusters, each almost independent of the others. For example, in manufacturing semiconductors, the more or less independent clusters include such things as wafer (die) fab, probe, assembly, and final test. My rule of starting from the end and working forward pertains to which ***cluster*** you address first. Within each cluster, you then apply the Theory of Constraints, and address the biggest bottleneck first.

FOCUSING ON CYCLE TIME

By continually compressing in this way—making each new problem top priority until it's been overcome—the production will be driven progressively to improve cycle time.

I've used a manufacturing cycle as illustration, but the same concept applies to how long it takes to design a new product, enter a new order into the system, or deliver freight to a customer—***anything*** that involves cycle time.

Certainly this isn't a new concept, yet managers continually overlook this simple manner of using time, rather than something else, such as money, to prioritize.

I don't believe in sitting around and wondering, "I have a thousand things that need improvement, what do I pick?" This active approach not only makes any job a lot easier, but gets past the time-wasting shuffle of prioritizing and theorizing. As soon as you say, "***This*** my most important problem of the moment," you are ready to move forward into action.

Addressing cycle time like this benefits you in two ways. It helps prioritize the problems that you need to work on. And it keeps you using the feedback loop as a learning tool.

Few organizations do this consistently—including National. I believe it's an area we all need to concentrate on because speed is among the most powerful of competitive weapons.

OPTIMIZING

With the right kinds of feedback loops—feedback designed to speed learning—any company continually improves its processes. As an illustration, consider the example from an earlier chapter of an organization's information systems.

Most companies already have fairly sophisticated systems for lot-tracking and the like. What I'm referring to goes beyond automation, to deal with the dynamic control of manufacturing. This is an area where information systems need to become much more widely used.

If a company wants to optimize what they build in their factory this month, how do they schedule it? They have a stack of customer orders, and they have a known amount of plant capacity. How can the two be matched in order to get the most out of the factory this month while achieving the best level of customer satisfaction?

They need to start by looking for the optimization principles. Remember that customer satisfaction is not what a company does for a customer but rather what the customer ***perceives*** has been received in terms of benefits.

The semiconductor industry has for the most part spent very little time thinking about optimization. But we're going to have to think about it, in order to shorten cycle times and improve learning.

Other industries have a comparable challenge. Imagine yourself as an executive with Chevron. One of your oil tankers ties up at the port of Oakland with a full load of crude. The crude goes through the refinery, and then must be distributed as fuel to the gas stations in the area.

The trucks begin to load; but how will you determine the most efficient way to get the oil from the refinery to the gas stations—which driver goes to which stations, in what order?

It sounds easy enough, but in fact turns out to be one of the most difficult, most intractable problems in applied mathematics. It belongs to an area of math that even has its own name—"linear programming and scheduling analysis." It's one of those tantalizing riddles that's easy to state, easy to understand, but present a Mount Everest of a challenge. And like the Everest summit, the real-world answers become lost in the clouds.

Managers in other industries have challenges very similar to this. Customers have different product demands, different time requirements, different packaging needs—a host of complexity. And the challenge is how to optimize the production schedule to both maximize the use of the resources, and keep the customers satisfied.

Although it's a problem that needs to be dealt with, few companies are addressing it adequately. At National, I'm convinced it's going to have to play a much bigger role in helping us optimize our return on investment.

WRAP UP

It's widely recognized today that companies, to stay competitive, need to shorten their development cycles, time to market, and the entire business process. At National, we need to get much better at this; the same is almost certainly true in your own organization.

To use time as a competitive weapon—

- increase your rate of learning
- use time to prioritize
- focus on improving cycle times
- use feedback loops
- use the principles of optimization.

In order to use time to prioritize, the method I recommend involves starting at the end of a process and working toward the beginning, continually making improvements, until a problem is encountered. Each time you encounter a problem, you then focus on the work cluster where this problem occurs, and (following the Theory of Constraints) determine which is the biggest bottleneck and address it first. You continue focusing like this, concentrating effort on each problem in turn until it has been cleared up.

The words we need to carve over the doorways of business for the 90s and beyond are "Speed is life."

16

Made to Measure

"What you measure is what you get... You can manage by what you measure."

Gil Amelio joined Rockwell in 1983 to solve the problems of one failing division—a $100 million operation, almost insignificant within the structure of the mega-giant $13 billion parent company. Yet he not only rescued the division but was eventually asked to give presentations to fellow general managers on his transformation process and his methods of doing financial analysis.

The programs he created in that small division were so powerful that they were ultimately adopted by the entire corporation. Consultant Mike Townsend describes the difference this way—

> *The earlier Rockwell was built to some extent on an operations-centric 'plant manager' business model. After Gil had transformed SPD, other division presidents within the company came to be held much more accountable by Rockwell management for things like customer responsiveness, marketing, financial acumen, and strategic planning—things that Gil's success had amply demonstrated the value of.*

If you have the right tools and use them to ask the right questions, Gil says, "you really can change the world."

★★★★★

At National, Gil Amelio started talking about measurement, most people thought he must be thinking about manufacturing. But to the team in corporate communications, it was a familiar theme.

National's Corporate Communications director, Mary Ann Phillips, had been running various kinds of surveys for several years to check on how National was perceived by customers and the public, and tracking the results of the regular **Wall Street Journal** *"Corporate Report Card" surveys, in which Journal readers consistently reported a familiarity with National but just as consistently ranked the company well below average in the reputation category.*

Looking back, Mary Ann believes a large part of the problem lay with the reluctance of management to accept the measurements. Several times the executive management team at National was presented with the results of the internal and Journal surveys. But they always seemed more concerned with whether the results were reliable than with what new programs or actions might be called for.

National had also been having surveys done by Griggs Anderson Research, of Portland, Oregon. Like the **Wall Street Journal** *samplings, these surveys time after time found that National was well known in the industry, but comparatively low in terms of respect.*

Finally one of the Griggs Anderson team said to Mary Ann, "We've done a lot of surveys for National over the last several years, and we always get the same results—when are you guys going to change!"

Sometimes the challenge lies not just in conducting the measurements, but in believing and acting on the truths the data reveals.

ESTABLISHING METRICS AT NATIONAL SEMICONDUCTOR

One of the consistent elements I've discovered in troubled companies is that when people are asked what success looks like, they don't know, they can't tell you. Most people have only a gut feeling instead of a well thought through, quantified answer. What I look for is an answer that says, "Success in our company means 20 percent return on equity," or "Success means 42 percent per year gross profit," or something equivalent. And the measure doesn't have to be financial—it can be in terms of gaining market share, innovation, customer awards, or the like.

It's always been startling to me that people in a company are willing to work like Trojans even though no one has told them how they'll know whether they've been successful.

I work with a consultant who, after listening to the problems and the assignment asks, "how will you decide if I've succeeded for you?" She doesn't start a project until she knows how a client will measure the success of the effort.

In addition, she usually has her own list of measurements that support her personal definition of success. But when the assignment is completed, she reports back on what she has accomplished according to *my* measurement terms so I know clearly what has been achieved, and what hasn't. I admire the fact that she enjoys consistent success in this direct way. Her style is useful because it nudges the people she supports to be specific about what they want and how they'll know when it has been achieved.

I urge you to be clear on how you will be measured and how you will measure others before starting any project or program. In my view, there's no better way to set yourself or someone else up for success.

Corporate success is fueled by dedicated, motivated people. But even the most determined workers can't help make the business a success unless they can be continually aware of how well their operation is performing, in what ways it's falling down, and where the problems are.

Even successful companies, divisions, and plants are often victims of their own disinformation. The disconnection often results from not looking at the right numbers. I've seen management congratulate themselves on a job well done or continue blindly down the wrong paths because they didn't know what data to look at or didn't know how to understand it. National Semiconductor was a prime example—as I've made clear, the company was being led by hard-working, dedicated, well-intentioned people, who knew their company was in trouble but didn't know exactly why because they didn't understand exactly what financial information they needed to be looking at. This was no special illness at National, but a common disease afflicting many companies.

The needed information comes from measuring performances and processes. The challenge—no surprise—lies in knowing what to measure and what the reasonable goals are.

When I arrived at National Semiconductor and began to study the financials, the picture I saw was, to put it politely, disheartening. Some of the key factors are depicted in the following charts.

Two of these plots show widely used measures, and the other two, though specific to the semiconductor industry, are measures of manufacturing. You don't have to know the detailed numbers to see that in every one of these important metrics, National was trailing the industry average. It was from these measures, among others, that I began to understand how uncompetitive this company really was.

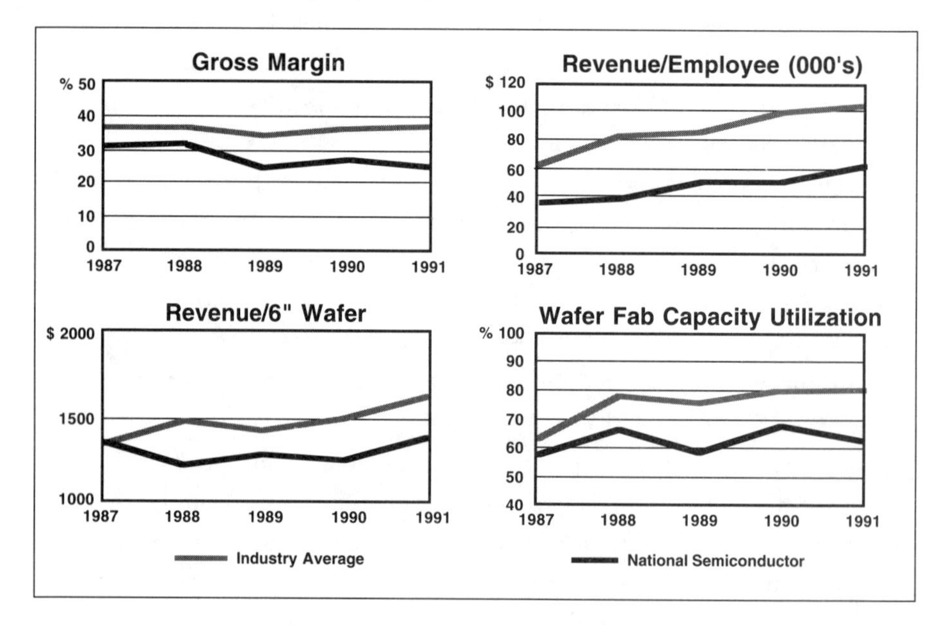

NSC Performance Trends, 1987-91.

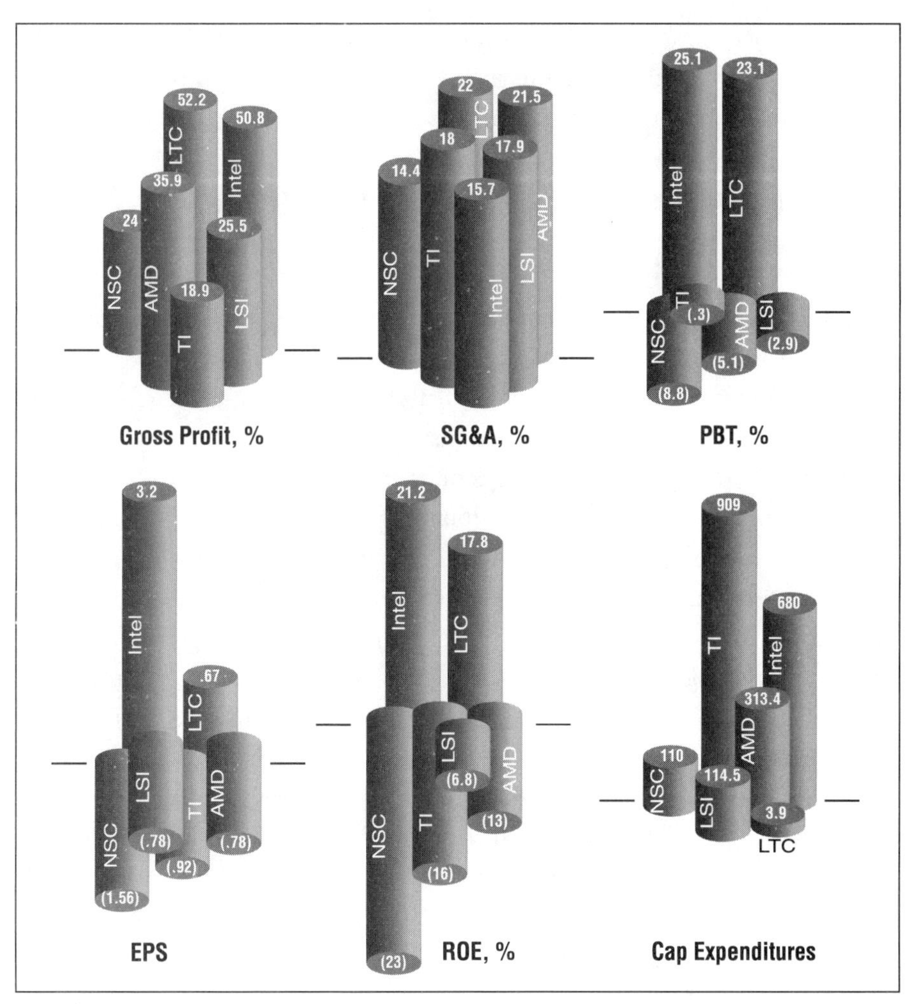

Performance of NSC compared to several major semiconductor industry leaders. Data is 1991 for NSC, 1990 for all others.

COMPETITIVE COMPARISONS

Most of the financial measures at National pointed to trouble—in particular, gross profit and return on investment. The company was also suffering under a heavy R&D burden, heavy in the sense that National, in the previous five years, had spent over a billion dollars on R&D, yet at the end of the period saw a gross profit that had declined. The real purpose of R&D is to drive up gross profit dollars; if it's not increasing, the R&D money is being spent unwisely. By this measure, National had spent a billion dollars and been left with precious little to show for it.

DEFINING SUCCESS IN TERMS OF INVESTMENT PERFORMANCE

The A. T. Kearney study mentioned earlier (in connection with span of control) also examined the issue of return on equity, and reported that top performing companies had better return on equity than average—20 percent for the best vs. 12.7 percent for the rest. Though the time span of that study was 1970 to 1988, the conclusion is still valid today.

One of my early actions at National was to provide the management team a new definition of success—defining it in terms of investment performance. At the first Executive Off-site, I called on the company to achieve a return on equity of 20 percent, and then set the team thinking about how we would manage our assets, how we would manage our financing and capitalization, how we would manage our business, in order to drive for that result.

As of the end of our fiscal year in May, 1994, gross profit was 41 percent—in two years, we had moved it up more than 17 percent. And return on shareholder equity had not just reached 20 percent—it had soared to 27 percent. We had come a long, long way—demonstrating how very much can be achieved by an intense focus on a specific goal.

A great many of the major, prestigious firms don't perform very well in return on equity—in the 80s, Standard and Poors companies averaged only about 12 percent—and yet return on equity is undeniably a key measure of financial success.

ROE is, of course, just one measure of performance. Depending on the situation, other factors may be given significant weight. These might include consistency of earnings, market segment leadership, the rate of revenue growth, and so forth. My choice of ROE for National was because the company had lost its credibility with respect to profits and asset management—hence, a high ROE would prove that we had understood the message and fixed the problem. In 1995, we're giving equal weight to ROE and revenue growth.

THE PROFIT PERFORMANCE MODEL

In determining what metrics to use, I have a cardinal rule: Measure the output, not the input. If you've settled on what your goals are going to be and what performance you're going to achieve, then you'll find it easy to decide on the right input.

Following this principle, you arrive at a list of the most valuable, the most essential, measurements for keeping track of how your company or operation is performing. Note that these apply to running a company of any size, and they apply equally to managing a product, department, division, plant, government agency, non-profit group, or virtually any other organization.

Including the ones I've already discussed, the list of crucial measurements looks like this—

- gross profits
- break-even
- investment performance
- profit on value added
- asset management.

These measurements will be covered in this and in the following chapter.

THE SUCCESS MODEL

One useful tool in helping people grasp the goals you're trying to set is one that I call the "Success Model," which establishes what you're shooting for, what you think your business ought to be able to do. It says, if we were running the company the way we should be running it, this is how the numbers and the ratios would look—15 percent per year growth, 40 percent gross profit, and so on.

"Success Model" is, if you will, a generic term that can take the form of a balance sheet, a traditional P&L, a "Spending P&L" (similar to a traditional one, but with the inventoried costs treated as period expenses), or any another quantified way of measuring performance.

The Success Model establishes a base. When you then take the reality of what you're actually doing today and compare it to the model, you may find that one part of the business is growing perfectly fine but not making any profits, while another may be making good profits but growing inadequately.

Based on what the Success Model reveals, decisions can then be made on which aspects of the business will be handled with band aids, and which need surgery or prosthetic limbs. For each item that's out of kilter with the model, I put a program in place.

In summary form, the detailed Success Model I presented at the first Executive Off-site looked like this—

NSC Financial Objectives

Gross profit	40%
Profit before tax	13%
Break-even sales	75%
R&D	11%
Return on Net Assets (RONA) (pretax)	33%
Return on Shareholder Equity (ROE)	20%

I believe in developing the Success Model early in the transformation process, as early as practical—and it's something I suggest be done with the entire management team. Like so many other aspects of transformation, the Success Model must not be a tablet handed down from the mount, but rather a guideline the team helps to develop. If they participate and believe in it, you will be able to count on their support.

I began to achieve my goals with some distinct measure of assurance after introducing the idea of the Success Model at Fairchild Camera and applying it rigorously. On moving to Rockwell I took it with me, and it was a major factor in the success we were able to achieve there.

However, that effort was hampered by an unusual problem which makes me smile to think back on it now. Somehow, despite my limited experience, I was able to recognize one easily overlooked difficulty, which was that the bookkeeping of the division followed the general methods used throughout the rest of the corporation—which had been created to serve the needs of a giant conglomerate that built spacecraft and aircraft components, and other things very different from semiconductors. Not surprisingly, the accounting system didn't work well for us, especially for things like determining product cost, and it became clear that the local accounting had to be redesigned to support our needs.

We made a lot of changes in how we accumulated costs, and in the process I had my financial group explode the P&L to a much greater level of detail. On each line I told them to show the name of the individual I had made responsible for that item—not an organization, but the specific person with the responsibility for ensuring this item would come out according to the Model.

At National, one of my first actions was to assign responsibility so that each person knew he or she had an "X" on the forehead marking responsibility for that line on the Success Model. We started at the top, establishing two major groups, each headed by a group president, who then established P&L responsibility within his product lines. (Some of this had existed before at National, but was not consistent throughout the company.)

We then took each of the Success Models and "parsed them down" through the organization, assigning individual responsibility for each item. For example, Dennis Samaritoni, who is in charge of procurement and facilities management, was advised that he would now be held responsible for meeting our Success Model at this level. Under him are people who look after each subelement of his total responsibility; thus, Jim Critzer has the "X" on his forehead for seeing that the purchasing department meets its own Success Model, which is a portion of Dennis' model. The sum of all of Dennis' direct reports adds up to his result.

The Success Model has probably been the most effective tool in focusing the National Semiconductor executive team and managers on the critical financial issues. I continually refer to it, they continually refer to it, and they use it within their own departments as well.

There are probably other organizational leaders who use this approach with the same intensity I do, as a way of focusing attention and getting buy-in, but I haven't run across them yet. My strong suggestion is that you make this practice into a management habit.

SUCCESS MODELS FOR YOUR COMPANY

Developing your own Success Models is a reasonably straight-forward process—

> Take one of your key financial statements—income statement, or balance sheet, for example.
>
> Benchmark each item on the statement: obtain market research data from your industry association and from your competitors as to what's average, and what the best companies are achieving.
>
> From the benchmark numbers, set a difficult but achievable target for your own company.

When you have done this for each key business parameter, you will have produced a model defining what success means to your organization. Then pass the challenge through successive levels of the company, so that each division, plant, group, and so on, develops its own set, each reflecting the goals of the next higher level.

PRODUCT LIFETIME COSTS AND REVENUES

Another financial tool that has proven highly valuable is a good one for people who, like me, appreciate simple rules of thumb or sometimes use the backs of envelopes for planning. This one is based on a rule of thumb I've used for many years.

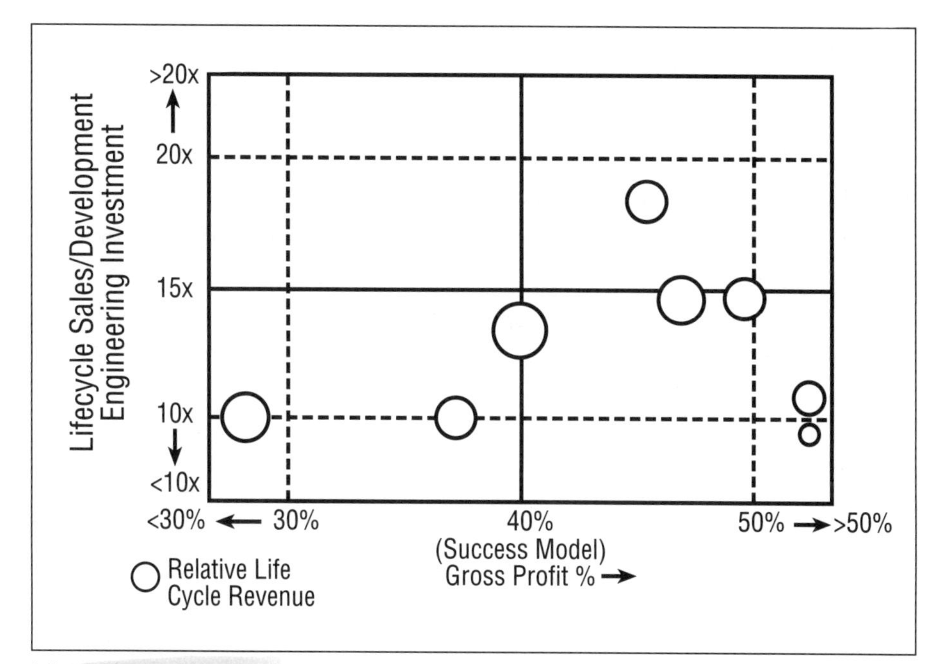

The Life Cycle Diagram—Product lifetime sales divided by development engineering costs, plotted against gross profit. The results give a good quick analysis of the R&D Return on Investment. (This model is limited in that it does not take into account the time value of money. However, if most of your products take about the same amount of time to bring to market, then the model is correct on a relative basis.)

The process for arriving at this chart begins with taking the life-cycle sales—the revenue you expect to derive during the entire lifetime of the product—and dividing this by the development engineering cost—what it will cost to develop the product. The result is then plotted against gross profit, giving what's sometimes called the "hurdle rate." In the semiconductor industry, a hurdle rate of 40 percent is frequently accepted as the go/no-go dividing line for determining whether a project makes sense and should go forward; you'll use a number appropriate to your own industry.

Next, apply my rule of thumb to determine whether the project in question is going to be a winner. If the result of the calculations is above forty percent with a multiple of fifteen, plot it in the upper right square; that's the outcome you want. (Unfortunately, before the transformation began at National, only a handful of products in our entire line met these criteria.)

Because these numbers vary somewhat from industry to industry, I suggest consulting industry associations and general industry knowledge to research the details appropriate to your situation.

Although at National we've now developed more sophisticated analytical techniques for studying these questions, in the early days I actually used this chart, which despite its simplicity embodies a lot more insight than meets the eye.

And the approach has great appeal because it's so marvelously easy: by making a few educated assumptions, you can get an answer on a paper napkin in the company cafeteria.

Take some of the products you're currently working on, plot them on this chart, and see where they fall. Even the process of drawing the graph is tremendously instructive and this exercise will determine very early on whether the product is likely to make any money for your company.

If you share this technique with your development engineers, they will rapidly become aware of how it can help them make the kind of pragmatic decisions that contribute to company wealth. (To get the denominator, the engineer must figure out the development cost—clearly, a worthwhile exercise. To get the numerator—the estimate of life-cycle sales—the engineer must go speak to marketing people, customers and the like—once again a very worthwhile process.) From experience I can predict that engineers will relate to this method very quickly.

IMPROVING YOUR R&D RETURN ON INVESTMENT

While we use return on investment to determine whether we're succeeding today (for the corporation as a whole and for tracking the success of each individual project), we use R&D ROI to determine whether we're succeeding in the efforts that will govern the company's future. Return on investment from R&D is perhaps the most valuable indicator of whether we're doing the right things with R&D, and doing them the right way.

The question then becomes, What can you do to get more "bang for the buck" from R&D?

Two of the efforts under the R&D ROI umbrella are—

- Phase Review Process: A tracking system which includes questions on the economics of what we're doing. Are we investing money in the right place? Are we going to get a payback from this project? The Phase Review process, like a warning sign advising of a washed-out bridge, enables us to terminate a project before we're in over our heads. My experience has been that what kills companies are the mistakes that go on year after year—a disaster you can avoid by recognizing and dealing with the problems in a timely fashion. The Phase Review process provides you a mechanism for doing this.

  I consider Phase Review to be an incredibly important and valuable tool, and I am sharing with you the guiding document we use at National. It's presented in Appendix D.

- Core technical competencies: This is in fact an ongoing focus on our incompetencies. The goal of this exercise is to be certain that we concentrate on the set of competencies at which we are presently among the best in the world. Leadership should never become sluggish about checking whether the company core technical skills and abilities are imbedded in all products.

  By succeeding in this step, we maximize the value to our customers and erect barriers to our competitors.

I strongly recommend further reading in this area. In particular, I suggest the works of C. K. Prahalad (who is a sometime lecturer for us) and coauthor Gary Hamel; see the End Notes for detailed references.[1]

MEASURING PEOPLE—NOT JUDGING THEM

We accept without flinching the idea of measuring things financial, but many are uncomfortable about whether measuring *people* is appropriate; maybe it's okay in the factory, but not with professionals and white-collar workers. Attitudes that evolved in the 1960s tell us we shouldn't judge others and should not have to be judged ourselves. As individuals we know what it feels like to be judged, so we hesitate to place others in that position. What's required is the ability to distinguish when we're judging versus when we're *measuring*.

One of the aspects of leadership I have found necessary to develop is the technique of measuring people. The key, it seems to me, is to measure people in terms of results, instead of trying to judge and control behaviors.

Consider a plant manager reporting to you—a traditional manager. Each month you could tell him how much new material to start in the line, and he'd set out to manufacture exactly as you had instructed. In a sense you have judged him less competent than yourself; in a standard supervisory style, you are controlling his actions and limiting his contribution.

A better way handles the situation with far less control: you say, "This is the output the company wants from your plant for the month and we'll be checking on those production numbers on Friday of each week at 4 p.m." Then you let *him* figure out how to achieve the goal so that his plant will be measured as successful. He'll have to determine how much raw material to buy, how to organize his labor, and the rest. Of course, you may also need to offer training, provide suggestions from other experienced people, and give him reassurance about risks and errors.

But most importantly, you have let him know how you will be measuring his success—you'll be measuring his output, and, as well, measuring how *efficiently* the factory is operating. You're also going to track costs, how many fixed assets he uses, how many square feet he takes up, how many employees he needs. You reassure him that you are going to measure these things together, and jointly plan how things can be improved in the future.

Sure, this approach puts a lot more burden on the plant manager. But of course, any plant manager worthy of the job looks on this as great news. People want to be measured in accomplishing what they've spent years training to do.

Provide that satisfaction and you nurture one of your company's most valuable assets—the brainpower of its people.

The old style of managing forced a plant manager with twenty years experience into a continual negotiation about how to get his job done. He felt robbed of his true self-worth and began to work in ways that management may have judged combative, negative, or objectionable. Judging people instead of measuring results is a confidence-killer. And as all enlightened people know, confidence is what makes the wheels turn to produce a profit.

Is the topic of judging people or measuring results just a new-age way of evading realistic management? Not in my view. However, to ensure that the new way really works, the essential ingredient is to select the right metrics—the right measurements, and then to be certain they're used consistently by everyone throughout the company.

Only when valid yardsticks are selected will employees be motivated to work and make decisions in ways that are both supportive and consistent with the company's vision and goals.

But in a few words, I've just described a very difficult task. Imagine trying to go around a large and complex company, selecting the right sets of tasks to measure, and then choosing the accurate measurements for each. In the words made famous by Winston Churchill, this is a blood, sweat, and tears process.

Yet in the final measure, without any exception, this is a search worth all the time and effort. In fact, if a company wants to stay in the game, choosing accurate measurements is *essential*.

In a broad sense, measuring offers a two-tier benefit. It provides information that's absolutely vital for running the operation successfully. This is the obvious part.

The second, hidden, benefit may be just as important: measuring is one of the most effective ways of communicating with people down to the grass-roots level—letting them know by what you measure, which aspects of their work they should view as the most important.

Of course, it follows that if the wrong things are measured, the results are more than merely wrong answers—people will unintentionally be encouraged toward the wrong kinds of behavior. Incorrect measurements, just as powerfully as inept rewards, will attract people toward the wrong kinds of behaviors. In some companies, sending confused signals like this goes on all the time. Over the years I've often witnessed confused decision making and wasteful behaviors that are clearly the result of inaccurate measurements or the absence of measurements.

A group at National was once about to kill an R&D project even though everyone seemed to agree it was a valuable program. Curious about the decision, I asked the manager for his thinking and discovered the company had been using a formula that saddled project managers with a disproportionate share of overhead. This particular manager had looked at the situation and said, "I can't afford this project in my new budget."

His decision appeared sound—he was adhering to a constraint the company imposed, and the numbers supported his conclusion. The accountants supplying him the data were applying the prescribed formulas in exactly the way they were supposed to, and the formulas were loading the project with a cost that made it look unreasonable—expenses disproportionate to the potential benefits.

But when the manager was asked, out of context of all those financial formulas, whether he thought the project would be beneficial to the company—he was certain it would be. Our people were being led to decisions that were actually contrary to the company's best interests.

You know as well as I do that this kind of confused decision-making takes place all the time. And yet, most of a company's outdated measurements are more forgivable than the ongoing efforts to cling to them!

Another example: a company has a chance to win a big order in Europe, involving the sale of a variety of items made by a number of different product lines. Everyone takes for granted that each of the product managers involved has to be able to say whether the deal makes sense. But that means any one of them could torpedo the whole thing if he or she doesn't like the pricing or the terms and conditions.

So the salesman in Europe sweats out trying to close the deal—he's on the telephone in a different time zone, trying to coordinate fifteen product managers from his own company to sign up for the program. That's no way to run a railroad. It's a straight roadbed to disaster.

I've often been heard to say that "setting measurements precedes managing—you cannot manage anything you can't measure." I'm convinced that the capturing of accurate measurements will start the transformation engines running.

WHAT YOU MEASURE IS WHAT YOU GET

When I arrived at Rockwell, the cycle time in our wafer fabrication operations was about eight weeks. The most appropriate term to describe an eight-week cycle time for that era's process technology is "abysmal."

For half a year, the responsible executives tried a variety of approaches to reduce the time, but nothing they could think of would help.

I proposed an experiment. We prepared a large chart of the cycle time—a big, multicolored, eye-catching affair—and we posted it in the fabrication area. It was very easy to read, and you couldn't be in the fab without noticing it. The chart prominently displayed a dotted line extending out into the future, showing the goal.

We didn't do anything else—just posted this chart, and kept the plot up to date.

It may seem simplistic but we are all merely human; like magic, cycle time started coming down. Not all the way, but enough to get the momentum going so the professionals could re-engage with the work force and start making some real progress.

It works. You *can* manage by what you measure.

Think about the following example and see if there isn't some place in your company that you could try it now. Every Monday morning, call one of your employees and ask very pleasantly something like, "How many units did we produce last week?" Or "How many calls did we process?" "What was the average time for xx?" Devote a few minutes to doing this every Monday, week after week.

At first the employee will be nervous about coming to work on Mondays because he knows your call is coming. Soon he gets used to preparing the information and gains confidence. And don't underestimate the extent of the pride he feels because he has mastered the situation by being prepared.

Has the behavior changed? Yes, of course it has. And, even more importantly, now that he has figured out this piece of information is valuable to management, communicating this information will become an established part of this worker's "tool box."

If you are consistent about asking for that same information, the production rate will probably go up, or the time will come down—it will be a win/win situation.

By asking people to report regularly about something specific, and thus telling them what to measure, their time and their thinking is redirected according to the metrics you've announced.

Of course, there's a caveat accompanying this method. Ask every Monday how many packages of copier paper your people used last week, and the number will probably start going down; morale may go down at the same time, as people begin to wonder whether the company is suddenly in such bad shape that the cost of supplies has become an important factor. When asked the wrong questions, people get focused on the wrong measurements, which practically guarantees that the important things are not receiving the attention they deserve.

The mere act of having people report information regularly is a strong and influential form of managing what people are doing—what they focus on.

On a regular basis and especially at the start of the transformation process, place a high priority on setting quality time aside in your schedule for thinking and discussing with your management team and with trusted consultants what should be measured, what you want reported. Place a new respect and trust on your measurements, not just on your management style.

BENCHMARKS

I've seen the activity of measurement-setting become just a useless and wasted exercise. The effort turns worthwhile if the manager has a way to know where there's a real need for improvement and whether improvements will be significant enough to make an impact.

We discussed benchmarks earlier, in connection with the critical business issues. Benchmarks are an important part of measuring, as well.

Fifteen years ago, five hundred defective parts per million was an acceptable level of quality for mass produced semiconductors. Today, numbers like that would force us out of business. For each item you're measuring, you need to be able to assign a quality judgment—is the performance great, adequate, or a disaster?

In Chapter 10, I suggested contacting other companies to obtain comparative information. The object is to benchmark the competition—setting up standards of bad, good, or world-class, based on what your competitors are doing. Take seriously the information about how you're performing in comparison to the competition.

Here are few examples of typical measurements and benchmarking:

- average gross profit in backlog.
- design-ins—the number of customers per month who are designing your product into theirs.
- accounting—How long does it take you to close the books? How many manual entries do you have to make to get the books closed? How much are you spending on audit fees?

I described earlier my effort in which I asked twenty companies for information to establish a benchmark, offering each a copy of the results, and found that they all cooperated.

Benchmarking requires some effort, but it's much easier to do than most people think.

WRAP UP

To keep you in touch with the vital aspects of what's really happening in the organization, few tools are more valuable than an effective use of metrics. Even more than that, metrics make it clear to those in the company what management considers important: What you measure is what you get.

I initially relied on gross margin and payroll turns (payroll divided by sales) when seeking to understand the situation at National.

I recommend you follow the principle, "Measure the output, not the input." In keeping with this guiding idea, I consider the following to be critical items for tracking the health and success of your operation—

- gross profits
- break-even

- investment performance
- profit on value added
- asset management.

For tracking investment performance, one measure I rely on is return on equity. While an ROE in the low teens is not uncommon, the best-performing companies frequently achieve an ROE of 20 percent. But revenue growth, as well as other factors such as earnings and market-segment leadership, are also important measures.

I recommend you and your management team together define a series of success models that spell out the financial goals you are trying to achieve. At National, we created a success model that set goals for gross profit, profit before tax, break-even, return on net assets, and return on equity. For your own model, find out what the average and the best values are in your industry, and what your competitors are achieving. Then set targets that are difficult but achievable.

A useful way of tracking the success of your R&D efforts is by measuring R&D return on investment. Two approaches I strongly endorse for improving the success of your R&D efforts are—

- the Phase Review process (detailed in Appendix D); and,
- arriving at a clear understanding of what constitute your core technical competencies.

Finally, remember that what you measure is what you get.

END NOTES

1. For further reading on core competencies, see "The Core Competence of the Corporation," *Harvard Business Review*, Vol. 90, No. 3 (May-June 1990), pp. 79-91.

Also see the book by these same two authors, *Competing for the Future: Breakthrough Strategies for Seizing Control of Your Industry and Creating the Markets of Tomorrow*, Harvard Business School Press, 1994.

17

From Red to Black

"Make gross profit the pivot point of your P&L.

Manage investing, not spending."

When Gil Amelio arrived at National, the executives expected the unexpected. Would he ask questions they had never been asked, look for information they had never considered important?

They would soon find that Gil's list of questions included these—

- *At the front end, what are the strengths and the weaknesses in strategic planning, product planning, and market development planning?*
- *What is the approach to cost accounting?*
- *How is the data collected? (Information Gil uses to gain insight on how the people have been looking at problems)*
- *How are the sales and marketing people establishing customer relationships, and what are they doing to maintain them?*
- *In engineering, what's the performance in getting new products to market?*

But most of all were the financials. Although he made it clear that each industry and each company is different, Gil nonetheless had a well-developed set of principles about what numbers most clearly show the true state of affairs.

The National executives and managers, who had been handling the company's financial matters for years, were about to learn some new ways of approaching the subject.

★★★★★

Everyone who has ever run a business or a business unit knows there are times when you want to shave unnecessary expenses to make the financial picture look better. Among Gil Amelio's favorite National Semiconductor stories is one that illustrates the point, yet he is often reluctant to tell it for fear he will be misunderstood; he prefaces the story with warnings that it shows a human side of one of his and the semiconductor industry's favorite people.

In the early days of National Semiconductor, when all the money they had was being put into research and new productive capacity, it was often touch and go as to whether the company had enough money to pay its salaries and other current expenses. During one business downturn, the situation became worse than usual, and Charlie Sporck decided they had to make some cutbacks. Among other items, he gave orders to stop cutting the grass around the headquarters buildings.

In high pressure situations, it helps if your people have the ability to defuse stress with a sense of humor. Luckily National has many people with that admirable balance; one was Bob Widlar, probably the best linear designer in the history of the semiconductor industry.

After a few weeks, the grass around the buildings had grown knee-deep. One day, Widlar brought a goat to work and tied it to a stake he drove in the middle of the lawn. The goat had a grass feast and everyone else had a laugh feast.

Even Charlie managed to throw back his head and guffaw; he then gave the nod to ignore the (admittedly minor) financial consequences, and resume the grass cutting.

In Gil's view, an ability to defuse stress with humor is not always a necessary quality for plain and simple management. But, he says, true leadership requires the ability to laugh—to find humor in stressful situations—and most important, to develop the ability to laugh at yourself.

WHAT I SAW AT THE INCEPTION

One of the questions I'm most often asked is about how I made the initial decisions on first taking over.

I've written in these pages about discovering that the company's financials didn't give insight into "the profit equation," which is usually the case when a company is in trouble.

The accounting practices were not providing the kind of information I needed to frame a valid picture of what shape the company was in and where the problems were.

People want to know what information I was looking for and what numbers I rely on in order to judge the health of a company.

One of the first things I look at is gross profit, as mentioned earlier, yet of all the divisions in National's two major corporate groups, and of all the thirty product lines in this company, only one of the executives and managers thought that gross profit was important enough to include as part of the initial presentation. In fact, when I first reviewed the financials, there was only that one business group in the entire company with the GP included among the figures.

On a financial break-out, cost of sales may have hundreds of items listed under it, which makes it easy for GP to get lost in the shuffle. National's management, from product managers to vice presidents, perceived gross profit as just another subtotal in the spending picture.

GROSS PROFIT: THE PIVOT POINT

For a manufacturing business, and often for other businesses as well, the gross profit number holds the chief clue in determining the overall health of a company, a product or product-line, or any other profit center. The financial spotlight belongs on gross profit, as the pivot point in the profit-and-loss statement. If you don't manage any other number in the P&L, you'd better manage gross profit.

And if you don't take anything else away from this book, you'll get more than your money's worth if you remember this one principle—

Make gross profit the pivot point of your P&L

After studying the industry and the company, after examining the established products and the products still in development, I arrived at an initial model income statement (the "Success Model") for National—

Net sales	100%
Cost of sales	60
Gross profit	40
Research & Development (R&D)	11
Marketing and Sales (M&S)	10
General & Administrative (G&A)	7
Profit before taxes	12

In corporate financial discussions, you sometimes hear the term "below the line." It's worth taking a moment to consider this term.

For some readers, the following might seem too familiar and too obvious to include here. Unfortunately, I continually discover that many experienced people in business have either missed this point completely, or missed the significant power of it. Certainly some of the managers at National Semiconductor were overlooking the impact of gross profit, otherwise they would have been showing that number in their income statements all along.

The "line" that below-the-line refers to is, of course and obviously, the gross profit line. For illustration purposes, we can write the above set of figures like this —

Net sales	100%
Less: Cost of sales	60
Gross profit	40
Below-the-line spending and profits	
R&D	11
M&S	10
G&A	7
Profit before taxes	12
Total	40

In other words, gross profit is what you have left over after taking out all the items that add up to the cost of sales—the manufacturing, labor, and overhead costs associated with production (sometimes translated into jargon as MLO). Gross profit is the center of the profit and loss.

The fact that the income statement shows a subtotal in the middle carries no significance to a lot of people. For those who persist in this mindset, all spending is equal. Wrong. All spending is *not* equal.

It's simple—if gross profit is zero or negative, then there is no money to invest in the below-the-line areas. These are what some people call "discretionary expenses," because in theory the manager can decide what to spend. But if you are without any GP, there's no discretion available: you cannot develop new products, you cannot get them into the market, you cannot advertise them, you can't earn any revenues from them. If you don't have any GP, you are in a death spiral.

Some managers focus on sales, others focus on profit before taxes. But the proper place for your focus needs to be on gross profit. Again, this may appear obvious, but I can give hundreds of examples where people don't get it or ignore it.

THE PROFIT EQUATION

The "Profit Equation" provides a way of depicting the influence of the various items involved in the gross profit calculation.

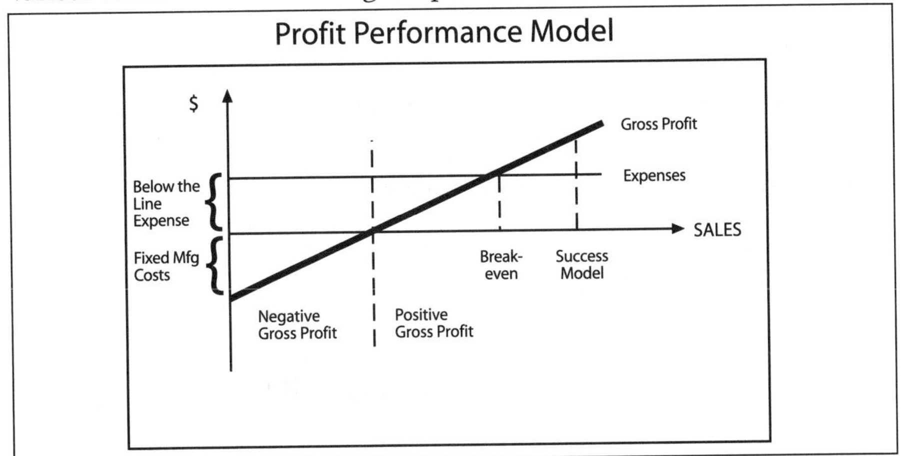

Profit Performance Model—The Profit Equation.

The chart plots expenses—the cost of doing business—against sales revenue.

Starting a business, or a new division, or a new product, means you begin at zero sales, with some fixed manufacturing expenses. In a major operation, just turning on the lights the first morning may cost you $1 million or $10 million—because the infrastructure must be in place. The people, the plant, the machinery—all of these constitute fixed manufacturing costs. The unpleasant reality is that you haven't yet sold anything and you're already in the hole.

As sales grow, and assuming price is greater than variable cost (as it must be if the product is to make any money), you begin to see some profit; on the chart, the gross profit line begins to move in the positive direction.

If everything goes according to your expectations, sales continue to grow and before long the two lines cross—you've reached break-even, the point at which gross profit covers all the expenses.

After break-even, if business continues to follow this pattern, before long you reach the "Success Model." It's what a business aims for—what you thought your business would be able to do.

Let's look at a concrete example. Suppose you're making a communication chip that you sell for $10. To make the chip entails a lot of factory costs—lights, taxes, water, corporate salaries, etc; these are overhead—the fixed costs. In addition there are the *variable* costs (I'm speaking here about the *volume* variable costs—not variable in the sense that accountants sometimes use the term, which generally means *time* variable).

The volume variable costs are those associated with making each chip—the silicon, the masks, the chemicals, the direct labor, etc.

Let's say the variable costs come to about $3 per chip. The $7 difference is then available as a "contribution" to eliminating the fixed overhead—often referred to as the "contribution margin." And when you sell enough of the chips, you can pay the fixed costs, and have some additional left over for profit.

These fundamentals aren't new to you, but perhaps this way of looking at them is.

Now let's try some variations.

LOWERING BREAK-EVEN

The lower break-even can be kept, the less your company will be exposed to risk and the more money you'll eventually make. Clearly, one way to reduce break-even is to reduce expenses. Too often, this is as far as some managers look. But you should also think about "pivoting the slope" of the gross profit line.[1] This can be done by improving the volume variable cost percentage—which is the extra cost it takes to make an additional unit. If you can reduce this value (which in the semiconductor industry is yield), the slope will increase. In most manufacturing businesses, changing the slope of the line is the most powerful item working towards improving profitability and reducing break-even costs—yet it often doesn't get the attention it deserves.

BREAK-EVEN AND THE EFFECT OF CREEPING EXPENSES

Suppose you're managing a division or a business that has been gradually growing, but recently has started losing money.

Let's take the same information we were just using, and plot it as dollars vs. time.

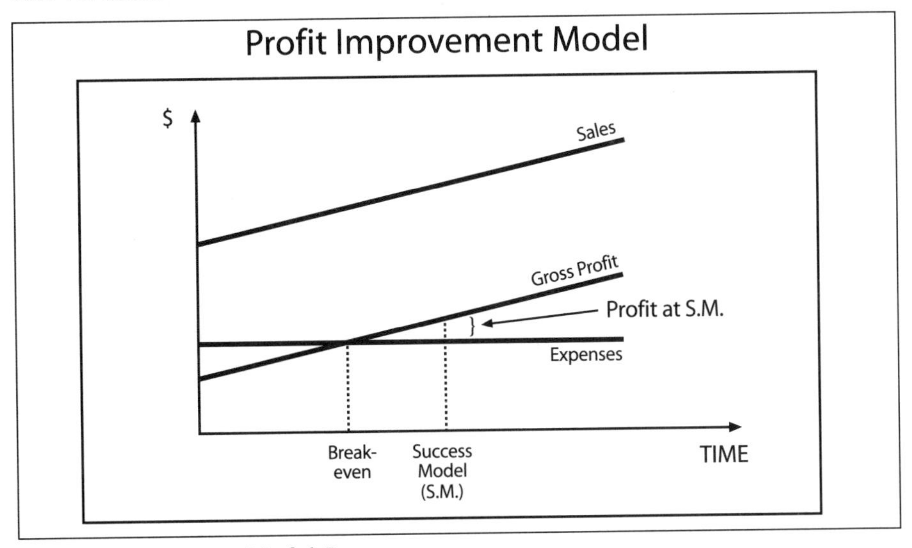

Profit Improvement Model-I.

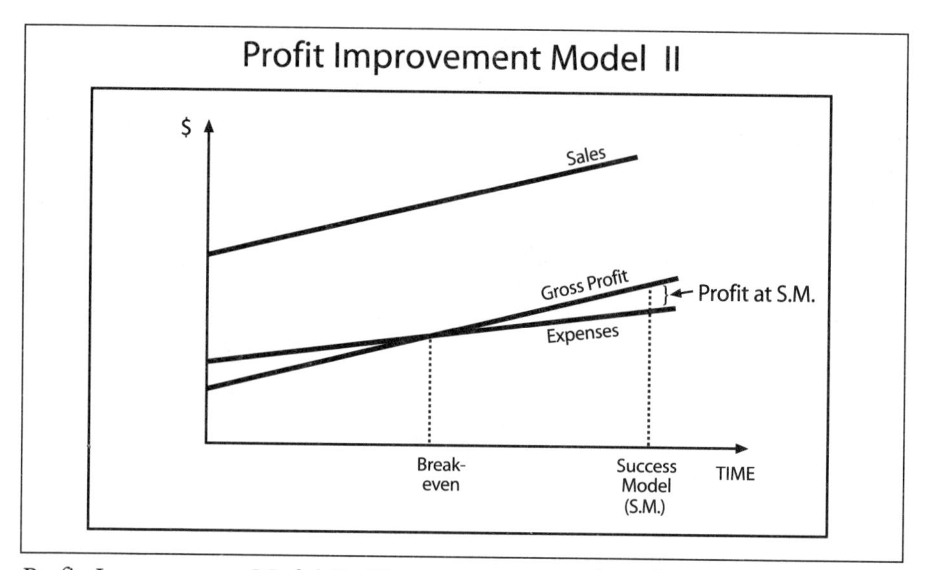

Profit Improvement Model-II. The consequences of not keeping expenses flat: Break-even occurs much later, and it takes much longer to achieve the same amount of gross profit dollars as in the previous chart.

The first chart depicts the situation we all envision: sales are rising, expenses are flat, and GP is rising. The second chart depicts what too often is actually happening: sales and GP are rising, but expenses are rising as well. Typically the company is being pressured to award the annual five percent raise, people feel that they can spend more because sales are increasing, taxes are going up, and so on. Soon you realize that this healthy flat line isn't flat after all—it's creeping upwards.

And when the expense line has a positive slope, even a gentle one, the break-even point moves out to the right—into the future.

If the slope of the expense line continues to increase, how fast does break-even move out? At the speed of light—faster than anything you've ever seen!

Holding the line on expenses always takes discipline, power, and courage. In my view, this is the single most important discipline you can acquire in the financial aspects of business—

Place a ceiling on below-the-line expenses and hold to it

LIVING UNDER A BURN RATE CEILING

Venture capital managers putting money into new businesses follow this principle by setting a value for what they vividly term the "burn-rate ceiling," which dictates to the people running the business that they cannot spend more than x dollars per month. Typically, the figure is based on allowing eighteen to twenty-four months for getting the product out, and they simply divide the amount of below-the-line financing by the number of months.

I suggest you do the same for your venture. In every troubled company I've ever seen, the slope of the expense line is positive—sometimes *very* positive. It's up to management to make sure it stays flat, or, at least, below the ceiling.

As for determining how much money you can reasonably spend on BTL costs, unfortunately there's no formula to fall back on—there are many variations depending on the type of business, how full the barrel is that holds cash reserves, and so forth.

But once having arrived at a decision, don't change it, don't let it creep. I urge you to develop the discipline to hold on until a change is clearly justified.

When the Success Model performance is achieved, the time has arrived to make a decision as to whether or not you want to increase the below-the-line spending.

In fact, this is the way to manage *all* investments in business: not as a steady-line progression—which doesn't give sufficient control and too easily, too often, leads to deep trouble—but rather as a series of stair steps. A simple but fundamental principle.

Again, it may sound obvious that you must hold BTL expenses flat, increasing only in stair-steps upon reaching goals—yet it appears that many, if not most, companies and managers don't do this.

BREAK-EVEN AS A PERCENTAGE OF SALES

Do you know what your break-even is *as a percentage of sales*? Is it going down every quarter? If not, what are you doing about it?

This concept is important because it captures both spending and margins. When margins are deteriorating, even if spending is kept flat, break-even is going up. A lot of people get blind-sided by this problem.

Tracking break-even as a percentage of sales is one way to stay out of that trouble. To control this number, you must manage *both* figures simultaneously—which keeps you from getting so fixated on one of them that you forget to pay attention to the other.

RETURN ON EQUITY AND RETURN ON NET ASSETS

Financial analysts typically rate a company in terms of earnings per share, but this isn't necessarily the best indicator of performance.[2] A better way of measuring business performance is return on equity (ROE)—how much money is being made, divided by shareholder equity. (In fact, some corporate leaders are critical of ROE, which is admittedly not a perfect measure; current thought in some quarters favors a new concept called "value management" as representing a still better way of measuring shareholder value. But for our purposes here, ROE is a more common and more readily available value.)

A widely accepted rule of thumb says that first tier companies should have a return on equity equal to or greater than 20 percent. Given the tax situation that most companies are in, and assuming a typical value of leverage, most companies need to have a return on net assets (RONA)[3] in the range of 20 percent to 30 percent to achieve this level of ROE.[4]

ASSET TURNS

Continuing to work backward with these calculations, the next question to ask is, "What profit before tax (PBT) percentage should I be able to achieve in this business?"

The PBT percentage is a value that everyone in business should have benchmarks for; let's say that in your business, it's reasonable to expect a value of about 11 percent.

Next, look at asset turns—an important value because it provides a measure of your capital productivity, and of your competitiveness. It's defined as—

Asset turns = Sales / Net assets.

What value of asset turns should you be aiming for? We can gain insight into this by using a pair of already-determined values —

RONA = (Profit before tax percentage) x (Asset turns); so,
Asset turns = RONA / Profit before tax percentage.

If PBT is 11 percent and RONA is around 33 percent, then net asset turns will be approximately three.

Three turns out to be a remarkably universal number. It not only applies to the semiconductor industry but works for many, if not most, manufacturing businesses as a measure of a healthy company.

Again, the significance of asset turns lies in its value as a very important indication of how competitive the company is in attracting capital. It provides an answer to the question, "For each $1 an investor places with us, how successful are we in using that to generate sales and profits?" And, as well, "How successful are we compared to the other companies the investor might put his $1 in?"

HUMAN ASSET PRODUCTIVITY

There are, I've come to believe, two simple metrics best used to measure productivity: in addition to capital productivity, discussed above—which has to do with the physical assets and physical productivity of the company—human asset productivity should also be measured. This value is obtained when sales are divided by labor costs, a measure of the people you have on the payroll—

Human asset productivity = Sales / Total cost of labor.

Note that for asset turns, sales are divided by net assets, and for human asset productivity, sales are divided by total labor costs. The first is a measure of physical things while the second is a measure of productivity of the people on the payroll.

What should this number be? For a typical manufacturing company, where gross profits are in the 40 to 60 percent range, a competitive value for this number turns out to be—surprise!—three.

The corporate finance people will tell you it should really be 3.156 or some such, but by simply remembering three, you will play safely in the right ball park. The game is about knowing whether your company is ready to compete or needs to put in a heavy bout of spring training before the playing season begins.

One of the goals that I set early on for National was to get both the capital productivity and human productivity numbers to three. As of early 1995, human asset productivity is still only at about 2.9. We're closing in on it, but we're not quite there. In any case, as is my practice with goals of this kind, we'll never manage to reach the goal because I always increase the target as we get near; I have already raised this goal to 3.3 so we will stay competitive on a global basis.

The issue of productivity is more fundamental to success than is widely recognized. When all the other competitive abilities are taken out of the equation, what's left is a question of which company is the most productive. Consider the way in which the personal computer is dramatically changing white-collar productivity. On occasion the evidence may seem contrary but, nonetheless, this is what's happening, and we're going to see over a relatively short period of time a quantum improvement in white-collar productivity.

Any company not keeping up will be left in the dust, and so we all need to figure this into the core of our transformation plans. Productivity during the next few years is going to be more important than ever.

HUMAN VS. CAPITAL PRODUCTIVITY

It's important to realize that there's an optimum balance between the two forms of productivity. If a company invests heavily in leading-edge equipment to increase the productivity of their people, *capital* productivity will necessarily go down.

Or the opposite choice can be made, which is what National had historically done—holding a very tight lid on capital equipment expenditures. So, what did the plant managers do when they were called on to increase production? They hired more people, which resulted in squeezing fixed asset productivity very high and pushing human asset productivity very low.

This is short-hand, not a science. But it's a valid and easy way to consider the balance between productivity and profitability. Continually ask, "What are the elements that drive each of these numbers?"

How are payroll turns and revenue per assets driven up? With better products! I call those "high protein products"—things people are willing to pay more for because they offer more value.

Value can also be increased in other ways than the value contained in the product. Perceived value resides in the many other benefits that are included as an intrinsic part of the product—service, quality, on-time delivery, short cycle times, short lead times, and so on.

That's the current productivity model we are using at National.

ASSET MANAGEMENT

Let's compare two companies. Recalling that return on net assets can be obtained as the product of net assets turns times profit before taxes (RONA = PBT * net asset turns) gives the following—

	PBT, %	**Net Asset Turns**	**RONA, %**
Company A	13.3	3.	40
Company B	20	2.	40

Which of these two companies is the stronger? Which would you rather be running?

The key to the answer lies in another question: Over which do you have more control—profits or asset management?

Profits are always subject to the threat of the market place—a competitor's new product or price reduction, a downturn in the economy. In a tight period, the company with better asset management will prove to be the stronger. This is a corollary to another belief I hold—that a company with high profits is at risk in a downturn. The company with higher net asset turns is in a better position.

Under pressure, Company A can afford to give up a few percentage points of profit. Under the same pressure, Company B will lose their profit position faster. If both companies get down to 10 percent PBT, Company A still has a 30 percent RONA, Company B only 20 percent.

Because so many external factors can influence profits, a manager has more direct control over asset turns than over profits. So my advice to the manager is—make as much profit as you can, but set your business up to have high asset turns.

The company with higher asset turns than its competitors will have a leading edge in its industry, because it will be less subject to the stresses of the economic cycle.

A CLOSER LOOK AT NET ASSETS

There's a tendency to treat net assets as a single item, when in fact it's made up of two components: fixed assets, and working capital. In National's business arena, because factories are so expensive, fixed assets account for about 85 percent of net assets. But in an industry like personal computer manufacturing, where the companies are largely assemblers, working capital dominates.

It's important to be clear about this distinction for your own company. If it is working-capital dominated, management can afford to be a little more relaxed about, for example, whether a new manufacturing machine will really pay for itself.

What's needed is to take net assets apart and decide which is the driver for you. The driver directs which of these items—fixed assets, or working capital—needs to dominate as a major focus of management attention.

MANAGING INVESTING VS. MANAGING SPENDING

Now consider another instance of the principle that "Not all spending is equal." Any expenditure on which a return is expected should be viewed as investing; any on which you don't expect a return, constitutes spending. You need to make clear distinction between expenditures of funds that create future value versus those that don't.

In order for transformation to get underway at National, I decided the Leading Change program was necessary for all employees. Necessary but expensive: it was budgeted on the order of 0.5 percent of sales—about $10 million.

Given the fact that at the time the company was losing its shirt, could hardly pay the rent, why in the world was I increasing the G&A spending? An armchair quarterback might have said, "That's nuts!" But my reason was sound: to make sure that National would have a future.

General & Administrative spending for such things as employee training typically impacts *future* gross profit. But I expected the Leading Change program to be so powerful that it would influence *current* gross profit. Although that would be judged very unusual for a G&A program, the assumption seemed justified to me, and indeed it proved to be a wise investment.

It's essential to review below-the-line items in terms of whether they are driving gross profit or not. To me, it seems bizarre that so few managers understand this, or understand it but fail to include it in their figuring and decision-making. Again, I urge you to learn the principle and make it part of your transformation arsenal:

Manage investing, not spending.

This also means that you can't "save your way into success." The temptation is to focus totally on managing spending, but in the long run deciding on where you're going to invest is much more important.

ALLOWING YOURSELF TO GET SUBJECTIVE

Before making an important decision—about launching a new product or project, creating a new workgroup, or almost anything else that involves a commitment of resources—once you have done the measurements, collected the benchmarks, run the numbers, and analyzed it all, there are three more questions you should ask yourself before arriving at a decision:

- Is it real?
- Is it worth the risk and effort?
- Can we win?

I first heard this list from Tracy O'Rourke when he was my boss at Rockwell; I've seen variations on the theme over the years, but the essential thrust is usually the same. These questions call on you to make subjective evaluation, based on your years of experience and your right-brain, intuitive, perception. They also give you a way of challenging yourself with a key question that's usually overlooked:

Am I managing risk... or avoiding it?

FOCUSING ON THE PROBLEMS

There's another aspect of spending that, once again, seems obvious but often gets lost in the shuffle. Suppose you find yourself approaching the end of a quarter and realize that the factory hasn't been able to meet production expectations—the numbers are going to fall a little short. If you were to follow the response suggested by the goat story, you might eliminate lawn cutting as well as slashing travel, R&D, and advertising. But if you respond by just tightening the screws on all expenses, you will be unable to focus management energy on the critical issue of investing and, in fact, merely discourage the organization. I put this in the penny-wise and a pound-foolish category of "cosmetic" management.

If you stop running ads, if you keep the sales force at home, how does that help you get your factory problem fixed? Of course it doesn't.

So it's not what the company spends, it's what the company earns that really matters. The central idea here is: If you have a problem, go to the source and attack that problem. It's always much better to focus on the real problem—low manufacturing yields or whatever—and take the heat if you don't make the numbers for the quarter. Management won't be happy about it, but they'll be happier in the long run because you'll get the real problem solved a lot faster.

This is a tough choice to make; here again the Winston Churchill quote is apt, "Courage is the understructure of all virtue."

ACT CONSERVATIVE FINANCIALLY AND AGGRESSIVE TECHNICALLY—NOT VICE VERSA

In my experience with large corporations, there's a very strong tendency for people to act conservatively on technical issues, and then take financial risks. And often these same people would describe their actions as financially conservative.

Nearly everyone from Polonius to Donald Trump spouts the traditional wisdom of being conservative in financial matters, while it appears to me they frequently make important decisions that are just the opposite.

In technical areas, I hear people say things like, "We're not going to use those new design rules, we're going to use the old design rules. We need to be safe." A manager who insists on using old design rules for creating a chip, increases the likelihood that the resulting product will function on the first try, but also increases the likelihood that it will not be performance-cost effective. The manager then attempts to overcome this drawback by forecasting huge market share and unrealistic yields.

Better to avoid this self-delusion, making the profit equation work with modest market forecasts and conservative yields by taking on the challenge of moving the technical state of the art forward. What I tell our engineers is that they know a whole lot more about the technology than they do about markets or manufacturing issues, and can therefore manage the rules better. (Which is probably why engineers so often get it wrong on their first try!)

Groucho Marx once advised a husband who wanted a happy marriage to learn to keep his mouth shut and his checkbook open. Transforming a company is not at all like transforming a marriage.

So be decisive about acting conservatively when it comes to finances and aggressively when it comes to technical matters. Take a close look at what you're doing rather than what you're saying.

WRAP UP

Below-the-line expenses have a tendency to creep upward; more realistically, they *will* creep upward unless you keep them under firm control. For companies in trouble, establishing a burn-rate ceiling that's flat over time, and then making absolutely certain that below-the-line expenses don't go over this limit, is absolutely essential.

The right way to manage expenses below the line is to set a ceiling and honor it. I suggest keeping to that ceiling until you reach your Success Model and are justified in setting a new one.

Especially in a capital-intensive field, you need to understand asset management and use it as a tool. If you are really demanding about asset management, and keep an "honest" set of books in terms of showing what your real cost of quality is, this will force you to focus on the right thing.

My short list of the critical financials and financial techniques, despite a bit of overlap, provides a focus of items that always deserve your keen attention—

- the Profit Equation
- the Success Model
- gross profit
- break-even
- investment performance
- R&D return on investment
- asset management
- return on equity and return on net assets
- asset turns
- human asset productivity

Finally, remember that gross profit is the pivot point of the P&L; without it you don't have a business. Pay more attention to gross profit than anything else on the income statement.

END NOTES

1. By definition, the slope of the Gross Profit line is obtained as a result of the calculation—

Gross Profit % = 1 - Cost of Sale %

Or, Gross Profit $ = Revenue $ - Cost of Sales $

2. An attractively clear explanation of this is given by Robert G. Hagstrom, Jr., in his book, *The Warren Buffet Way* (John Wiley & Sons, 1994)—"Buffet considers earnings per share a smokescreen. Since most companies retain a portion of their previous year's earnings as a way of increasing their equity base, he sees no reason to get excited about record earnings per share. There is nothing spectacular about a company that increases earnings per share by 10 percent if, at the same time, it is growing its equity by 10 percent. That's no different, he explains, from putting money in a savings account and letting the interest accumulate and compound." (p. 87.)

3. The relationship between return on equity and return on net assets is expressed by the formula—

ROE = RONA (1 - tax rate) * leverage

4. For most large companies, the tax rate is around 35 percent. Furthermore, most U.S. companies outside the high-tech industry carry a debt to equity (leverage) of around 30 percent; for these companies, a RONA in the low twenties is acceptable. In the case of National, where we have essentially no debt, RONA must be more like 30 percent. Typically, high-tech companies carry very low debt because of the risk/volatility of the business. At our best, National was achieving RONAs of 40 percent and more with some individual divisions in the sixties and seventies. If your company is above the nominal range of 20 to 30 percent and the market believes this is sustainable, you are considered to have a very attractive stock. The ability to sustain relies on having a clear differentiated, competitive advantage. A good example of this is Microsoft.

Afterword

As of the end of our fiscal year in May, 1995, National Semiconductor made approximately $330 million income before tax for the year, had a low debt-to-equity ratio and had over $450 million cash in the bank.

The company's long-term debt was about $85 million—not very much, given the capital-intensive nature of the industry and the almost $2.5 billion size of the company.

Shareholders' equity, which at the end of 1991 had stood at about 540 million dollars, four years later had increased by $900 million, to about $1.4 billion. An R&D center which had been sold in a lease-back arrangement when the company was strapped for cash had been repurchased. And the rate of new-product introductions was about twice what it had been.

For a company so recently on the verge of going out of business, these achievements were clearly impressive.

But National was still a long way from the company it needed to be and could be. In human terms, while they could point to significant achievements, they still had much to do toward establishing a common vision of what they wanted to become and providing adequate education and training to carry out this quest. In financial terms they also had much to be proud of, yet were still hard at work on getting the ratios right.

Amelio and the NSC management team had started with an emphasis on ROE, and pushed it from significantly negative territory to 27 percent. But (except for a company growing at greater than market rates) the financial community gives little credit for an ROE above 20 percent. So in 1994 National had faced the bizarre situation of actually needing to reduce ROE, by increasing the investment rate as a necessary prelude to targeting a greater rate of growth.

In order to underscore the need to shift emphasis from ROE to growth, Amelio announced the goal of achieving a balance between ROE and revenue growth, with a stretch goal of 20 percent ROE and 20 percent growth. During "Phase II" (see the following section), managers are being measured on the combination of ROE and growth, and Amelio has had the compensation scheme weighted to help drive that growth agenda.

By 1995, Gil considers the National Semiconductor to be "roaring down the runway and ready to climb to new levels."

LOOKING AHEAD: KEEPING THE PROCESS GOING

The End of Phase I, the Start of Phase II

The initial phase of transformation—which has been the subject of this book—was designed to carry National Semiconductor from being troubled to becoming healthy. When the company reached the Success Model that had been set in 1991, Amelio congratulated everyone, and announced the launching of Phase II, which he describes as "taking the company from healthy to great."

A NEW STRATEGY FOR PHASE II

In Phase I, you are coping with the challenges of a problem-driven company; Amelio refers to this as "the crisis phase." As you move into Phase II, the need is to adopt new approaches that will further transform the organization into an *opportunity*-driven company, a company capable in some way of transforming the world.[1]

In Amelio's view, that would mean making National into a company such that "every time anybody, anywhere, turns their desires into ones and zeroes or gets on the World Wide Web of the Internet, they will do it through a piece of National Semiconductor silicon."

A NEW LEADING CHANGE COURSE FOR PHASE II

After your company has once again become basically healthy and vital, you need to shift people's focus to become seekers of opportunity. Phase II reflects this essential difference.

The original version of Leading Change was, according to Amelio, the major factor in propelling the people of National to both a higher understanding and enthusiastic support of the new Vision and goals. For Phase II, the company developed a totally new Leading Change program to impart new values and teach new management skills.

Like the original version, Leading Change #2 will over time reach every National Semiconductor employee, from top management to the plant floor.

GETTING RID OF PROCESSES THAT DON'T ADD VALUE

In the same way that all cities have blighted areas that need to be cleaned up but are mostly ignored, all businesses have practices, procedures, requirements, and whole work groups that may have made sense when they were created but are no longer valuable.

We know this to be true, but as with the blighted neighborhoods, we rarely find the time, energy, and resources to undertake an effective cleanup.

So—a parting reminder: take a fresh and courageous look at what's not working and what's not needed in your organization. If you do this regularly, your company will be in a constant state of transformation.

Continual transformation is the way companies stay young, vital, and productive.

PROMOTING EMPOWERMENT

In Phase I, you promote empowerment... yet it's a time when people are more focused on the attitudes and skills needed for transformation. While you teach and promote empowerment, and eagerly search for signs that it's beginning to happen, the truth is that Phase I progress only takes place when the leader is teacher, coach, and captain of the cheering squad, providing guidance and direction setting.

In Phase II, employee empowerment becomes critical. The leader puts his attention back where it's most needed—on long-term strategy. In this phase, there's far less top-down direction setting, and it becomes essential for employees to accept the challenge of empowerment.

Gil says—

> You've got to get people passionate about this so they're willing to commit the enormous energy it takes to keep transformation going.

... and Good Luck

What worked at National Semiconductor can work for you, too, in your own company, agency, plant, division, workgroup, or other organizational unit.

It can be an energizing, exciting and most rewarding voyage. And it should be.

Bill Simon and Gil Amelio wish you success.

> The tendency is for managers to focus only on the problems—that's human nature. If it's all you do, it's guaranteed the business will under-perform. Instead, look for the opportunities.

> Success does not exist within any transformation plan—success is won by people perceptive enough and determined enough to design a transformation plan that will work, and follow it.

END NOTES

1. As a good source of ideas for Phase II planning, Gil Amelio recommends *Competing for the Future*, by Hamel and Prahalad.

APPENDIX A

National Semiconductor Financials

BALANCE SHEET & INCOME STATEMENT BY FISCAL YEAR (FY)

$M	FY61	FY62	FY63	FY64	FY65	FY66	FY67	FY68	FY69	FY70
INCOME STATEMENT										
NET SALES	3.0	3.6	3.8	4.7	5.3	3.1	7.2	11.0	22.9	41.8
COST OF SALES	2.1	2.5	2.5	3.0	3.6	2.2	5.5	7.1	17.1	31.6
GROSS PROFIT	0.9	1.1	1.3	1.7	1.7	0.9	1.8	4.0	5.8	10.2
R&D	0.0	0.0	0.0	0.0	0.0	0.0	0.0	0.0	0.0	0.0
SELLING, G&A	0.8	1.0	1.2	1.3	0.7	2.4	3.0	4.0	7.1	
RESTRUCTURING	0.0	0.0	0.0	0.0	0.0	0.0	0.0	0.0	0.0	0.0
OPERATING INCOME	0.1	0.3	0.3	0.4	0.4	0.2	(0.6)	1.0	1.8	3.2
INTEREST (EXP)/INC	(0.1)	(0.1)	(0.1)	(0.1)	(0.1)	(0.1)	(0.1)	(0.0)	(0.1)	(0.5)
INCOME BEFORE TAX	(0.0)	0.1	0.2	0.3	0.2	0.0	(0.8)	0.9	1.6	2.7
TAXES	0.0	0.0	0.0	0.0	0.0	0.0	1.4	0.0	0.2	0.8
GAIN/LOSS OF DISCON-TINUED OPERATION	0.0	0.0	0.0	0.0	0.0	0.0	0.0	0.0	0.0	0.0
NET INCOME	(0.0)	0.0	0.2	0.3	0.2	(0.0)	(2.2)	0.9	1.5	1.9
EARNINGS PER SHARE:PRIMARY	$0.06	$0.22	$0.25	$0.43	$0.29	$0.08	$0.00	$0.55	$0.90	$1.13

$M	FY61	FY62	FY63	FY64	FY65	FY66	FY67	FY68	FY69	FY70
BALANCE SHEET										
ASSETS										
CASH & SECURITIES	0.0	0.0	0.2	0.0	0.0	0.7	0.1	0.2	0.5	0.6
ACCOUNTS RECEIVABLE	0.5	0.6	0.8	1.1	1.2	1.6	1.4	2.9	3.4	5.9
INVENTORIES	0.7	0.9	1.0	1.2	1.7	2.1	0.9	1.4	2.0	5.4
OTHER ASSETS	0.0	0.0	0.0	0.0	0.0	0.0	0.0	0.0	0.0	0.3
TOTAL CURRENT ASSETS	1.2	1.6	2.0	2.3	3.0	4.4	2.4	4.5	6.0	12.2
NET PP&E	0.5	0.7	0.8	0.9	1.5	1.6	1.6	2.4	5.5	11.2
OTHER ASSETS	0.0	0.0	0.1	0.1	0.2	0.2	0.0	0.0	0.0	0.0
TOTAL ASSETS	1.8	2.3	2.9	3.4	4.7	6.1	4.1	6.9	11.5	23.4
LIABILITIES & EQUITY										
CURRENT LIABILITIES	0.7	1.1	1.2	1.5	2.0	1.8	1.8	2.1	4.3	9.4
LONG TERM DEBT	1.0	1.0	0.9	1.1	1.2	1.3	1.4	1.2	1.1	4.9
OTHER LIABILITIES	0.0	0.0	0.0	0.0	0.0	0.0	0.0	0.0	0.0	0.8
TOTAL LIABILITIES	1.8	2.2	2.0	2.6	3.2	3.1	3.3	3.3	5.4	15.0
SHAREHOLDERS' EQUITY	0.0	0.1	0.9	0.8	1.6	3.1	0.9	3.7	6.1	8.4
TOTAL LIABILITIES & EQUITY	1.8	2.3	2.9	3.4	4.7	6.2	4.1	6.9	11.5	23.4
NET ASSETS FOR RONA COMPUTATION	1.0	1.3	1.7	2.2	3.1	2.9	3.3	5.6	9.7	
RETURN ON NET ASSETS (PRETAX)	26%	23%	27%	17%	5%	(21%)	30%	32%	33%	
RETURN ON SHARE-HOLDER EQUITY	69%	18%	39%	14%	(1%)	(251%)	24%	24%	23%	

BALANCE SHEET & INCOME STATEMENT BY FISCAL YEAR (FY)

$M	FY71	FY72	FY73	FY74	FY75	FY76	FY77	FY78	FY79	FY80
INCOME STATEMENT										
NET SALES	38.1	59.8	99.0	213.4	235.5	325.1	338.7	437.2	651.9	910.1
COST OF SALES	28.7	45.7	77.4	158.1	150.5	218.9	232.7	293.8	446.2	618.7
GROSS PROFIT	9.4	14.1	21.6	55.3	84.9	106.2	106.0	143.5	205.7	291.4
R&D	0.0	0.0	0.0	0.0	20.7	24.9	31.3	42.9	67.8	80.1
SELLING, G&A	10.1	14.4	25.7	32.8	46.9	46.9	57.8	83.5	122.5	
RESTRUCTURING	0.0	0.0	0.0	0.0	0.0	0.0	0.0	0.0	0.0	0.0
OPERATING INCOME	2.5	4.1	7.2	29.6	31.4	34.4	27.8	42.8	54.3	88.8
INTEREST (EXP)/INC	(0.6)	(0.4)	(0.4)	(0.9)	(2.0)	(1.1)	(1.7)	(1.9)	(5.5)	(11.1)
INCOME BEFORE TAX	1.8	3.6	6.8	28.7	29.4	33.4	26.1	40.8	48.8	77.7
TAXES	0.7	1.6	3.0	12.3	12.6	14.4	11.5	18.2	22.0	34.4
GAIN/LOSS OF DISCONTINUED OPERATION	0.0	0.0	0.0	0.0	0.0	0.0	(7.3)	(2.6)	0.6	1.7
NET INCOME	1.1	2.0	3.7	16.4	16.7	19.0	7.3	20.0	27.4	45.0
EARNINGS PER SHARE: PRIMARY	$0.32	$0.56	$0.32	$1.33	$1.34	$1.44	$0.78	$1.72	$1.72	$2.58

BALANCE SHEET										
ASSETS										
CASH & SECURITIES	1.0	0.4	0.9	2.2	4.2	8.2	4.8	4.9	8.3	7.5
ACCOUNTS RECEIVABLE	5.5	9.8	18.2	35.1	37.0	52.4	61.6	79.6	105.0	173.1
INVENTORIES	4.7	7.7	12.6	23.0	43.1	55.1	66.7	91.7	125.4	170.3
OTHER ASSETS	0.8	0.2	0.5	0.9	0.7	1.5	2.8	4.7	5.7	53.1
TOTAL CURRENT ASSETS	12.1	18.2	32.1	61.2	85.0	117.3	135.9	181.0	244.4	404.0
NET PP&E	9.7	13.2	20.7	38.4	49.7	50.3	73.3	94.0	135.0	205.8
OTHER ASSETS	0.0	0.5	0.6	0.8	0.7	3.4	3.7	3.9	5.8	5.7
TOTAL ASSETS	21.8	31.8	53.4	100.4	135.4	171.0	212.9	278.9	385.3	615.5
LIABILITIES & EQUITY										
CURRENT LIABILITIES	5.3	11.1	18.1	33.6	39.5	47.7	67.4	103.2	148.4	234.7
LONG TERM DEBT	5.3	5.4	3.8	9.7	13.5	2.6	2.2	1.5	27.0	96.8
OTHER LIABILITIES	1.6	2.7	4.2	10.0	14.1	26.8	36.2	43.0	41.9	87.0
TOTAL LIABILITIES	12.2	19.2	26.1	53.3	67.1	77.1	105.8	147.7	217.3	418.4
SHAREHOLDERS' EQUITY	9.5	12.6	27.2	47.1	68.3	93.8	107.0	131.2	168.0	197.0
TOTAL LIABILITIES & EQUITY	21.8	31.8	53.4	100.4	135.4	171.0	212.9	278.9	385.3	615.5
NET ASSETS FOR RONA COMPUTATION	10.3	15.7	23.9	42.4	66.1	82.9	96.3	116.1	157.2	236.5
RETURN ON NET ASSETS (PRETAX)	24%	26%	30%	70%	48%	42%	29%	37%	35%	38%
RETURN ON SHARE-HOLDER EQUITY	12%	16%	14%	35%	25%	20%	7%	15%	16%	23%

BALANCE SHEET & INCOME STATEMENT BY FISCAL YEAR (FY)

$M	FY81	FY82	FY83	FY84	FY85	FY86	FY87	FY88	FY89	FY90
INCOME STATEMENT										
NET SALES	1,110.1	1,104.1	1,210.5	1,655.1	1,787.5	1,478.1	1,867.9	2,469.7	1,647.9	1,675.0
COST OF SALES	763.1	846.9	928.3	1,146.3	1,258.5	1,096.9	1,319.1	1,732.9	1,280.3	1,251.1
GROSS PROFIT	347.0	257.2	282.2	508.8	529.0	381.2	548.8	736.8	367.6	423.9
R&D	117.7	109.1	114.7	158.5	204.6	222.4	218.9	280.2	264.8	252.4
SELLING, G&A	171.1	159.8	167.2	247.6	264.9	276.5	310.3	378.0	236.2	224.3
RESTRUCTURING	0.0	0.0	9.0	0.0	0.0	0.0	15.0	0.0	53.6	(8.0)
OPERATING INCOME	58.2	(11.7)	(8.7)	102.7	59.5	(117.7)	4.6	78.6	(187.0)	(44.8)
INTEREST (EXP)/INC	(14.1)	(5.6)	(14.1)	(1.3)	(7.7)	(16.1)	(13.9)	(3.8)	(11.5)	12.4
INCOME BEFORE TAX	44.1	(17.3)	(22.8)	101.4	51.8	(133.8)	(9.3)	74.8	(198.5)	(32.4)
TAXES	(8.7)	(6.6)	(8.7)	37.5	8.6	8.9	15.3	12.1	7.0	(3.1)
GAIN/LOSS OF DISCON- TINUED OPERATION	(0.3)	0.0	0.0	0.0	0.0	0.0	0.0	0.0	182.3	4.3
NET INCOME	52.5	(10.7)	(14.1)	63.9	43.2	(142.7)	(24.6)	62.7	(23.2)	(25.0)
EARNINGS PER SHARE:										
PRIMARY	$2.39	($0.46)	($0.20)	$0.75	$0.49	($1.10)	($0.36)	$0.48	($0.31)	($0.35)

BALANCE SHEET										
<u>ASSETS</u>										
CASH & SECURITIES	10.6	11.0	7.0	27.8	12.2	21.8	189.1	147.2	228.0	128.7
ACCOUNTS RECEIVABLE	193.6	189.9	191.0	264.4	244.7	220.8	269.2	204.9	194.2	211.4
INVENTORIES	170.5	177.5	174.0	223.1	223.4	205.8	270.1	227.6	218.9	220.0
OTHER ASSETS	57.2	60.6	67.2	84.3	128.1	101.0	96.5	197.6	37.3	64.9
TOTAL CURRENT ASSETS	431.9	439.0	439.2	599.6	608.4	549.4	824.9	777.3	678.4	625.0
NET PP&E	309.8	332.1	353.4	510.3	757.9	705.9	633.9	622.4	696.5	702.2
OTHER ASSETS	12.0	14.0	54.8	46.1	44.2	40.1	40.6	104.2	41.2	50.4
TOTAL ASSETS	753.7	785.1	847.4	1,156.0	1,410.5	1,295.4	1,499.4	1,503.9	1,416.1	1,377.6
<u>LIABILITIES & EQUITY</u>										
CURRENT LIABILITIES	238.6	286.9	282.5	404.2	375.8	353.4	473.5	405.3	448.8	426.7
LONG TERM DEBT	67.0	76.1	149.0	24.2	225.9	123.4	35.7	36.9	52.2	64.2
OTHER LIABILITIES	114.8	85.4	80.1	108.5	127.6	101.6	99.1	48.1	66.6	69.9
TOTAL LIABILITIES	420.4	448.4	511.6	536.9	729.3	578.4	608.3	490.3	567.6	560.8
SHAREHOLDERS' EQUITY	333.3	336.7	335.8	619.1	681.2	717.0	891.1	1,013.6	848.5	816.8
TOTAL LIABILITIES & EQUITY	753.7	785.1	847.4	1,156.0	1,410.5	1,295.4	1,499.4	1,503.9	1,416.1	1,377.6
NET ASSETS FOR RONA COMPUTATION	288.2	395.8	439.8	546.7	755.2	856.8	778.2	820.5	788.0	712.5
RETURN ON NET ASSETS (PRETAX)	20%	(3%)	(2%)	19%	8%	(14%)	0%	10%	(24%)	(6%)
RETURN ON SHARE-HOLDER EQUITY	16%	(3%)	(4%)	10%	6%	(20%)	(3%)	6%	(3%)	(3%)

BALANCE SHEET & INCOME STATEMENT BY FISCAL YEAR (FY)

$M	FY91	FY92	FY93	FY94	FY95
INCOME STATEMENT					
NET SALES	1,701.8	1,717.5	2,013.7	2,295.3	2,379.4
COST OF SALES	1,294.3	1,247.5	1,379.6	1,333.7	1,380.4
GROSS PROFIT	407.5	470.0	634.1	961.6	999.0
R&D	198.6	192.1	202.3	257.9	281.6
SELLING, G&A 2	41.9	251.0	284.8	411.3	402.9
RESTRUCTURING	119.6	149.3	0.0	0.0	0.0
OPERATING INCOME	(152.6)	(122.4)	147.0	292.4	314.5
INTEREST (EXP)/INC	3.6	5.4	2.9	11.0	14.6
INCOME BEFORE TAX	(149.0)	(117.0)	149.9	303.4	329.1
TAXES	1.3	3.1	19.6	44.4	64.9
GAIN/LOSS OF DISCONTINUED OPERATION	(1.1)	0.0	0.0	0.0	0.0
NET INCOME	(151.4)	(120.1)	130.3	259.0	264.2
EARNINGS PER SHARE: PRIMARY	($1.57)	($1.28)	$0.98	$2.00	$2.00

BALANCE SHEET					
ASSETS					
CASH & SECURITIES	192.5	159.2	331.8	466.8	467.4
ACCOUNTS RECEIVABLE	200.3	194.5	271.5	288.1	318.0
INVENTORIES	188.4	207.8	189.6	212.7	263.0
OTHER ASSETS	31.4	33.5	49.4	48.8	129.9
TOTAL CURRENT ASSETS	612.6	595.0	842.3	1,016.4	1,178.3
NET PP&E	527.4	518.7	577.4	668.0	962.4
OTHER ASSETS	50.7	35.2	56.8	63.3	95.0
TOTAL ASSETS	1,190.7	1,148.9	1,476.5	1,747.7	2,235.7
LIABILITIES & EQUITY					
CURRENT LIABILITIES	416.5	473.0	505.7	577.5	685.9
LONG TERM DEBT	19.9	33.9	37.3	14.5	82.5
OTHER LIABILITIES	96.0	102.6	96.1	50.1	60.6
TOTAL LIABILITIES	532.4	609.5	639.1	642.1	829.0
SHAREHOLDERS' EQUITY	658.3	539.4	837.4	1,105.6	1,406.7
TOTAL LIABILITIES & EQUITY	1,190.7	1,148.9	1,476.5	1,747.7	2,235.7
NET ASSETS FOR RONA COMPUTATION	579.2	449.9	478.5	598.1	837.6
RETURN ON NET ASSETS (PRETAX)	(26%)	(27%)	31%	49%	38%
RETURN ON SHARE-HOLDER EQUITY	(23%)	(22%)	19%	27%	21%

APPENDIX B

Leading Change: Schedule for the 5-Day Program

The early series of Leading Change, designed for VPs, GMs, and middle managers, typically followed this schedule:

Sunday	
3:00	Welcome/ Introductions
3:30	National time line
7:30	The new Vision—presented by a corporate vice president or other high-ranking executive

Monday	*Leading Corporate transformation*
8:00	Outcomes and Agenda
8:10	Transformation at General Electric—Bob Miles
10:30	Visioning Exercise
1:00	The 7S Model—Lincoln Electric Case
2:30	Rockwell SPD Case—Video
5:30	Reflections

Tuesday	*Achieving Financial Success*
8:40	Financial Success at National—Don Macleod or other executive
1:00	Powerhouse Computer case
3:00	National strategy; Issues & Opportunities
4:45	Physical team building activity
Wednesday	*Steps to Organizational Excellence*
8:30	Tools for Managing Personal Dynamics—Cynthia Scott
10:45	Organizational Change Tools: Initiatives for system change at National
1:00	Set-up for Vision workshops
2:00	Exploring the Vision in depth
4:00	Report & dialogue on the Vision in depth
7:30	What will it take to implement the Vision
8:30	Report & dialogue on implementing the Vision
Thursday	*Exploring the Vision*
9:00	Your role in Leading Change: Personal visions; How to begin
11:00	Reports; First Steps
1:00	Review Gil's expectations, set ground rules for Gil's appearance and the presentations to Gil
2:00	Prepare presentations
4:30	Presentation dry run. Feedback; ownership
7:30	Polish presentations
Friday	*Committing to Change*
8:30	Outcomes/Agenda
9:00	Presentations to Gil Amelio
10:45	Dialogue with Gil
1:00	De-brief; Communicating the Vision
3:00	Evaluation/Closure
3:15	Adjourn

APPENDIX C

Gil's Guidelines for General Managers

In his 1993 seminar at the Stanford Business School, Gil Amelio explained the origin of what has come to be called "Gil's Guidelines."

"In 1988 I went to Dallas to try to fix the telecommunications business that Rockwell had at the time, and it ultimately turned out to be a very successful enterprise. But I suddenly had a whole new group of managers reporting to me who didn't really know me very well, and we'd get in these business reviews and I would start commenting with what they began referring to as my 'one-liners.'

"And pretty soon they began to see a pattern to these, and they said, 'Why don't you write them down?' which sounded like a good idea. I like keeping things short, and the result was this list of five things that I think GMs need to keep in mind when they go about transforming a business."

In the document now used at National, the five original items are fleshed out with subpoints that clarify the meaning and method.

CREATE A VISION BASED ON INTERNAL VALUES AND ON DELIVERING VALUE TO THE CUSTOMER, AND SET THE INITIAL VECTOR

- Create a clear vision of the future; invite employees to discover the vision and contribute to its definition; define the corporate character; honor the past
- Think of business as a value delivery system
- Focus on the intersection of what customers value and your core competencies
- Define success clearly; identify critical success factors: the CBIs
- To gain momentum, take bold initial steps and communicate to all employees the direction you must move to achieve success
- Seek/identify/realize sustainable competitive advantages/ differentiation
- Keep the vision global; think in terms of global markets
- Realize you cannot be #1 in everything, but you must be #1 in something
- Product/service leadership is critical.

BUILD AN ORGANIZATIONAL FRAMEWORK WHICH REINFORCES SHARED VALUES AND TEAM BEHAVIOR

- Use a self-consistent framework such as the 7S model to guide and prioritize effort ("organizational excellence")
- Design the organizational structure on the basis of a fluid, multidimensional model. Minimize preferred axes. Encourage teams. Deemphasize status
- Demonstrate the art of leadership
  - ▲ Personal attributes
    - – be a role model, not a victim,

- know yourself,
- strive for balance: short-term vs. long-term, incremental vs. quantum, vertical vs. horizontal, etc.,
- know how to deal with ambiguity,
- get out of your comfort zone,
- understand how to steadily raise the bar,
- pursue lifelong learning,
- be a good communicator,
- practice situational leadership,
- acknowledge mistakes and correct them quickly.

▲ Set a goal of seeing that every manager

1) has a clear understanding of the vision

2) has the necessary skills, knowledge and training

3) has the right real time decision support information

4) is empowered to act

▲ Empower employees using the six step model;

▲ De-bureaucratize—delegate authority and responsibility to the lowest practical levels; treat people as our greatest asset: educated, informed, empowered, accountable, and rewarded;

▲ Recognize/reward/promote—you cannot acknowledge achievement too much;

▲ Communicate/train/develop—encourage individual renewal;

▲ Recognize that sometimes the process is more important than just getting results: the Welch matrix;

▲ Selecting and hiring—attributes to look for in rank order.

1) a good person—attitude and values

2) smart

3) experience

4) a sense of humor

5) works hard

6) intuition

ESTABLISH EXCELLENCE AS THE STANDARD

- Operational excellence is the #1 strategic weapon. Design-in/build-in quality in everything we do—internal and external
- Benchmark: measure performance against the toughest competitor
- Understand, develop and continuously improve business system processes; simpler is usually better
- Encourage innovation and creativity
- Insist on congruence of functional goals; demand discipline and require attention to detail; use the "raise-the-bar" process to redirect non-performers
- Maintain an external orientation. Build relationships/partnerships with customers, suppliers, collaborators, fellow employees, and other National divisions to gain buy-in. Treat the customer like your best friend. Achieve customer delight, gain commitment and intimacy.

COMPRESS THE CYCLES OF BUSINESS ACTIVITY

- Use time as a competitive weapon: shorten the cycle time of everything—target the weaknesses uncovered by this process
- Identify/implement/strengthen learning loops—become a learning organization (including time horizon impact)
- Make continuous improvement a way of life. Continuous improvement means continuous change; make every manager an agent of change—"you drive change or change drives you"

- Facilitate intra-business communications/information flow—avoid suboptimization
- Prepare well-defined, detailed schedules and make meeting schedules and commitments sacred.

MANAGE INVESTING VERSUS SPENDING

- Measure/track everything that's important—in real time. "You cannot manage what you can't measure"
- Use models and ratios liberally to focus action—but remember, the P&L is a lagging indicator
- Act financially conservative but technically aggressive (not vice-versa)
- Focus on gross profit, investment performance, asset management, productivity, and liquidity and cash
- Track the break-even point—quickly correct drift from established goals
- Beware of cutting costs in one area to make up for a problem somewhere else
- Develop and be prepared to use contingency plans.

APPENDIX D

New Product Phase Review System

Objective: Minimize time-to-market, while meeting needs of the customer, meeting the market window and achieving revenue goals to ensure the business success of the product.

PHASE I—NEW PRODUCT IDEAS

- Basic description of whole product, including an application example, typical customer profile, estimated market size, market window, potential competitors, barriers & possible intellectual property.
- What is the Value Proposition? (See back page)
- Develop a Preliminary Product Business Case &
- Product Requirement Specification (one page each)
- Discovery Team is formed
- Discovery Team Leader begins Project
- Documentation Notebook & entry into R&D Cost Tracking System

PHASE II—DISCOVERY PROCESS

- Understand potential customers, competitors, whole product definition and the technologies required
- Approval to proceed with development of New Product Execution Plan (NPEP)
- NPEP Team formed
- Additional time, resources & funding approved
- Update R&D Cost Tracking System

PHASE III—DEFINITION PROCESS

- Refinement of the Discovery Process
- Develop complete New Product Execution Plan (NPEP) (see next page) consisting of:
- Product Business Case (PBC)
- Product Requirements Specification (PRS)
- Product Development Plan (PDP)
- Product Quality Plan (PQP)
- Product Launch Plan (PLP)

PHASE III REVIEW

- Are necessary resources & funding available?
- Contract review (if required)
- Are plans aligned with SBPs?
- Update R&D Cost Tracking System

PHASE IV—PRODUCT DEVELOPMENT

- IC/PCB Design Review
- Software Architecture & Design
- H/W & S/W test development & methodology
- Test software
- Mechanical Package Design
- Collateral Materials Development
- Tape ready for mask shop
- Design & Process Failure Mode & Effect Analysis (FMEA)
- Patent plan to Legal Department

PHASE IV REVIEW

- Are PBC & PLP still valid?
- Update R&D Cost Tracking System
- Approval to begin mask making & fab
- Approval for tools & materials for prototype build
- Approval to proceed with software code & test

PHASE V—FAB, ASSEMBLY & SOFTWARE CODING

- Packaged Alpha silicon
- Fully assembled prototype PCB
- Fully assembled prototype H/W & Packaging
- Coded & unit tested S/W modules
- H/W diagnostic primitives/test vectors
- MARCOM package

H/W PROTO TESTING, S/W INTEGRATION BUILDS

- Debugged hardware platforms
- Compiled, executable software

INTEGRATION, DESIGN VALIDATION, MANUFACTURABILITY & INITIAL CHARACTERIZATION

- Characterized silicon & H/W platforms PMS Code A
- Mask work registration
- Tested software
- Bug report lists
- Technical Publications second draft
- Complete manufacturing documentation package

ALPHA SITE (OPTIONAL FOR ICS)

- Packaged Alpha silicon
- Characterized PCB and/or system H/W
- Tested S/W, including bug list
- Non-Disclosure Agreements (NDA)
- Preliminary customer documentation set
- Alpha Site agreement signed by customer
- Alpha Site Reports, including corrective action plans

PHASE V REVIEW

- Are PBC and PRS still valid?
- Update R&D Cost Tracking System

- Release to manufacture pilot build & testing
- Reliability & Quality
- Approval for Beta Site sampling
- PMS Code B

PHASE VI FINAL CHARACTERIZATION

- Manufacturable, qualified & characterized devices & H/W platforms
- Customer documentation packages

BETA SITES

- Packaged Beta silicon
- Test equipment & procedures
- Data sheets & manuals
- User information/errata
- Beta S/W & bug lists
- Characterized PCB
- Beta agreements signed by customers
- Verification of design, fixes
- Beta Site reports, including corrective action plans

PHASE VI REVIEW

- Are the PBC & PLP still valid?
- Updated R&D Cost Tracking System
- Approval to begin manufacturing
- PMS Code P/C/R

PHASE VII REVIEW

- Released product to manufacturing
- Update R&D Cost Tracking System
- Is the Product Launch Plan (PLP) on schedule?
- Is the Product Business Case (PBC) still valid?
- Is the Whole Product Support in place?

PHASE VII REVIEW PROCESS IMPROVEMENT INPUTS

- Actual performance against plan (NPEP) Significant process events
- Current actual market conditions

OUTPUTS

- Revised product process map
- Annotated NPEP guide for future product development
- Updated Project Notebook
- Update R&D Cost Tracking System

BENEFITS OF NPPRS

- Early involvement of customers
- Encourages use of empowered cross-functional teams, with responsibility for the project
- Ensure winning, lower risk new products
- Assures plans & documentation are consistent with company strategy
- Facilitates management & visibility of numerous projects
- Provides motivation for up-front research & planning

PROJECT DOCUMENTATION

- Collect all project information, documents and records:
- Team Charter
- Meeting Minutes
- Schedules
- Actual Results
- Plan Changes & Reasons for Change
- Phase Review Minutes & Actions
- Organized by Phase
- Becomes Project Archive
- Team Leader Responsibility
- Information entry into the R&D Cost Tracking System

RESOURCING & FUNDING

- Phase I Review
- Seed funding approved for further investigation
- Phase II Review
- Resourcing & funding for Discovery Team
- Phase III Review
- Approval of NPEP is the authorization for the NPEP Team to proceed, with resourcing & funding approved as stated in the NPEP

TEAMS

- Discovery Team
- Phase Review I thru Phase Review II
- NPEP Team
- Phase Review II thru Phase Review VIII

APPENDIX E

Strategic Planning at National Semiconductor

The procedures for strategic planning are set forth in an extensive document called the "National Planning System." The following is a summary of some highlights.

NATIONAL PLANNING SYSTEM OVERVIEW

Competition for the future is competition to create and dominate emerging opportunities. The National Planning System (NPS) is the process by which National's businesses and other organizations make the decisions that will determine their future.

Process: NPS strives to develop strategies and business plans to create quality and service that customers can see through a coordinated company-wide planning effort. It precipitates, integrates, and implements incremental improvements. It also facilitates identification of breakthrough opportunities and improvements during the creative process. There are two principle elements to the NPS: Strategic Business Plan-I (SBP-I) and Strategic Business Plan-II (SBP-II). The synchronization of all planning activity is accomplished by adhering to a company-wide planning calendar. Thus the output of one organization becomes the input for others ultimately resulting in Product Line and Support Group Plans.

The NPS is formally launched by the president/CEO who delivers a Prolog message to the management committee, which includes the group presidents. The Prolog sets expectations covering areas of emphasis and financial targets, at the corporate level. Each layer of management divides the objectives, personalizes the Prolog, and parses its pertinent expectations downward. During the Dialog stage, planning teams interact with their management to better understand objectives, to evaluate strategic options, to decide their recommended strategies, and to document their plans. In the Postlog stage the management committee reviews the plans, makes adjustments, and communicates the results back to the business unit managers.

SBP-I: The principle activity of NPS is the creation of Strategic Business Plan I (SBP-I). SPB-I is a relatively long-term plan at five years, or longer if desired. The primary emphasis is on strategic thinking and, particularly, creative thinking. The key inputs to SBP-I are customer and end-user needs, as well as the identification of new competitive spaces. A limited set of financial data is prepared primarily covering P&L, RONA, and Shareholder Value.

SBP-II: SBP-II is a subsequent effort that emphasizes the short-term. It focuses on specific first year financial plans and the activities to implement the strategies developed in SBP-I. The financial data for the first year by quarter covers the P&L, Regional P&L, RONA, and Spending. Though focused on the first year, it is also five years in scope in order to show a realistic connection with SBP-I.

WHO DOES PLANNING?

Four major entities do planning: Strategic Market Segments (SMSs), Product Lines, Support Groups, and Operations.

SMS: The SMS Strategic Business Plans (SMS SBP-I) provide long-term, generally five years, strategic direction throughout the company on behalf of their market segments. SMSs prepare a perspective of customers and competition unique to their segment as well as strategies to create and dominate emerging opportunities. Strategic Action Statements (SASs) are delivered from SMSs to all their constituent Product Lines and Support Groups at the "hand-off" following the SMS SBP-I. They clearly state the requirements from each Product Line or Support Group to fulfill the SMS strategy. SMSs also typically convey a high-end Product Migration Plan (PMP) to these Product Lines. By accepting the SASs and PMPs from an SMS, Product Lines and Support Groups agree to deploy resources to fulfill these requirements. SMSs are dependent on their constituent Product Lines for actualization.

Product Lines: The Product Line is the primary entity that has the resources to satisfy the customer. In a reinforcing relationship, SMSs and Product Lines reciprocally contribute their needs as internal customers to Support Groups to be addressed in their next round of plans. Each Product Line and SMS indicates its technology requirements and shows the R&D spending it feels will be necessary to obtain it. The regional Business Divisions (RBD's) of the International Business Group (IBG) participate on the planning teams of SMSs and Product Lines. The plans of SMSs and Product Lines are deployed within the RBDs through this participation. The Product Line considers all products including multi-market, not just those that are oriented toward a single market segment.

Support Groups: Support Groups are actually internal forms of suppliers. The plans of Support Groups make a particular contribution in the consideration of company capabilities. There are Technology Support Groups covering the wafer fabrication processes, packaging, and design tools. There are also Support Groups covering Human Resources, Quality/Reliability, Information Systems, Value Delivery, Materials/Logistics, Facilities/Environmental, and Corporate/Employee Communications. Support Groups are reviewed by appropriate Corporate Councils which are composed of key members of Divisions. The Council members take the messages from the Support Groups back to their organizations to ensure these Plans are embedded in their own Division's Product Lines and Operations organizations.

Operations: Operations organizations are also internal forms of suppliers. Operations prepares an SBP-I; however, their primary effort is an SPB-II using the five-year demand from the Product Lines' SBP-I's. This calibrates the reality of the company's ability to product these Product Line projections and to take advantage of new market opportunities.

EXPECTATIONS: Product Group Planning Teams

The guidelines described below are primarily intended for SMS and Product Group organizations. However, most other organizations preparing an SBP-I should find them helpful. The expectations for SMS and Product Group organizations are organized in eleven categories, which should provide the structure of the SBP-I document. Appendices may be added, but the SPB-I should stand by itself without extensive supporting appendices or other documentation. Note: For Continuous Improvement evaluation purposes, each of the 11 SBP-I categories is assigned a numerical value, shown in [ ] , totaling to 100. Evaluation will be done strictly on the basis of the planning document.

These guidelines represent the basic corporate requirements. Your management may add additional requirements, but may not subtract from these requirements. These guidelines provide basic expectations, rather than detailed prescriptions, exercises, and forms to fill in. The basic expectation is that planners will use good judgement, and that those who need help will ask for it. A wide variety of assistance and tools is available in both personal and written form; contact Strategic Planning and Market Development (SPMD). In very few instances, certain pieces of information are requested in specific formats; please observe these few clearly indicated mandatory formatting requirements.

1. *Who Did it?* [2]

   List those who contributed to its preparation (i.e., the team members). The team leader will be identified, and his or her VP/GM will literally sign-off on the plan, thus indicating agreement with the plan and willingness to support it.

2. *Reconciliations* [4]

   List (1) any previous planning commitments the team may have made (SBP-I, SBP-II, New Product Phase Review, etc.) that are no longer operative, or in which the underlying assumptions have materially changed; (2) All Prolog guidelines and whether the SBP-I contemplates meeting or missing those guidelines; and (3) (for organizations expected to contribute to an SMS) all SMS Strategic Action Statements from the "hand-offs" for which the organization is responsible (wholly or in part) and whether the SBP-I contemplates meeting or missing those requirements.

3. *Objectives of the Business* [10]

   Define the organization's scope and state what the organization expects to accomplish over the plan period. Ordinarily this will consist of basic financial objectives and market share and position objectives. It may also consist of the following objectives: Specific product, technology development, value-chain, regional, key account, manufacturing volume, industry image, objectives linked to other organizations' objectives (especially SMS's), etc. All subsequent aspects of the SBP-I should be directly traceable to one or more of these objectives, or they are otherwise strong candidates for rejection.

4. *Customers and Markets* [12]

   Divide the addressed market into relatively homogeneous customer groups and demonstrate an in-depth, multidimensional understanding of each group. The dimensions of this understanding will include: (1) end-user needs and trends; (2) National customer needs and trends, with the needs articulated to show relative importance; and (3) target customer set (the small number of customers, by name, who are expected to generate 50 percent or more of the organization's sales). The SBP-I will also segment the addressed market by major application, and for each such application (1) articulate trends (including technology, economics), (2) analyze the value chain(s) in effect and likely to develop, and (3) forecast market segment size (stating source of estimates). Market size forecasts will be expressed as a range corresponding to a stated confidence.

5. *Competitors* [12]

Competitors are particularly dynamic and difficult to predict so significant effort should be devoted to understanding competition in both a present and future perspective. Demonstrate that the planning team has an in-depth, multidimensional understanding of each major current competitor, and a good grasp of new competitors likely to emerge over the plan period. Indirect competition (e.g., from technology substitution) will be included. For each market segment (see #4 above), the top three competitors and their current and projected market share will be identified, as will the value advantages that account for each competitor's market share, projected over the plan period. Anything else about any competitor that accounts for it being a major threat will be noted. The organization will list what it will do to counter each competitor's threat.

6. *Strategic Plan* [15]

State the organization's strategic plan, which will consist of three parts: a strategic alternative analysis, Strategic Action Statements (SAS's), and a Product Migration Plan (PMP). In the first of these, the planning team will explain the alternative strategies considered and the basic rationale for what was selected. The second and third of these, together, will portray what the organization is going to do to meet its objectives, given its assessment of "current reality" plus objectives (#3), customers (#4), and competitors (#5). The SAS's will, wherever possible, be articulated in terms of measurable outcomes and specific timeframes. Topics to be considered include R&D and capital investments, core competence leverage, new technology developments, strategic partnerships, new regional thrusts, changes in the value chain, changes in sales/distribution strategies, what's required to "leapfrog" the competition's value advantages, anticipated competitor reactions, special handling of key customers, etc. The PMP will depict proposed product launches for a minimum of the first three years of the plan period; for each proposed new product, four things will be stated: (1) what fiscal quarter the product will be launched; (2) its planned initial ASP; (3) its expected ROI over its lifetime: and (4) its essential value advantage. The quality of the organization's strategic plan will be judged on the basis of (1) how credibly the proposed actions, including new products, support the fulfillment of the objectives, and (2) how complete and robust is the depiction of milestones and PMP.

7. *Key Assumptions* [5]

   Provide a thoughtful listing of the assumptions on which the plan is based. "Thoughtful" means a good-faith attempt to identify any assumption that, if subsequently not realized, could have a meaningful effect on the outcome of the plan's implementation. The assumptions will, wherever possible, be articulated in terms of measurable effects and specific timeframes. To facilitate future reviews of progress, it is mandatory that planners display on one timeline graphic the key assumptions made.

8. *Financial Summary* [8]

   Include the National SBP-I Finance Schedules. These depictions will be based on expected ("most likely") outcomes related to the range of results expressed in the risk analysis (#9, below).

9. *Risk Analysis and Range of Results* [10]

   List the risks to the plan's expected outcomes (#8, above). It will also list opportunities now captured in the expected outcomes. It will provide a broad-brush contingency plan for each major risk. It will recast the sales and PBT projections as a range of results around the expected outcomes, based on these risks and opportunities and on the range of market forecast shown in #4.

10. *Stretch Opportunities* [7]

    Identify additional investments beyond those contemplated in the rest of the plan that are believed by the planning team to hold substantial incremental opportunities for the organization. Note: These are different from the "opportunities" identified in #9, above, which, if they materialize, can be captured with little or no additional investment. Each identified opportunity will be reviewed/articulated in condensed/encapsulated form according to the applicable portions of #4, #5, #6, #8 and #9, above.

11. *Narrative Summary* [5]

Provide a narrative summary, signed by the author, from three to five pages in length that answers six specific questions: (1) What is/will be the size and structure of the market application segments this plan addresses? (2) What are/will be the key customer values that apply in each application segment? (3) Who are/will be the top three competitors in each application segment, what are their value advantage(s), and how will these be addressed by National? (4) What are the SAS's and the PMP? In summary, what are you going to do to WIN? (5) What financial results (sales, gross margin, PBT, RONA) are expected? (6) What are the risks, opportunities and range of results? The overall responsibility of the planning team in this section is to persuade a naturally skeptical reader that it will win.

EXPECTATIONS: PLAN REVIEWERS

It is the responsibility of those who review SBP-I plans to ensure that (1) the planning team has done the best possible job of analyzing current realty and likely futures; (2) the best strategy has been selected from among the available options; (3) the planning team has faithfully adhered to Prolog and SMS constraints and requirements; (4) the planning team has learned as much as possible from the experience; and (5) the plan itself is adequately documented. The plan is likely to be solid if it encompasses all of the items below which should serve as a checklist of review points.

1. *Plan Document*

Does the document lend itself to understanding the plan, without benefit of oral presentation?

2. *Who Did It?*

   Is the team named, the team leader identified, and the team leader's manager's signature affixed?

3. *Reconciliations*

   Agrees with the Prolog? Agrees with SMS requirements? Covers changes to previous planning commitments?

4. *Objectives of the Business*

   Includes sales, gross margin, PBT, RONA objectives over the plan period? Includes market share objectives over the plan period? Do these objectives justify continuing the business?

5. *Customers and Markets*

   Demonstrates (preferably superior) knowledge of customers? Focuses on customer value? Identifies a key customer group, with plans to nurture? Provides an applications-based market segmentation? Shows in-depth understanding of the value chain? Shows independent opinion about reliability of market size forecasts? Are forecasts intelligently enveloped, rather than CAGR-extrapolations? Is all the above articulated over the five-year plan period?

6. *Competitors*

   Demonstrates (preferably superior) knowledge of competitors? Attempts to predict future entry? Deals with threat of technological substitution? Top three competitors for each market segment identified, with current and projected market share? Are these market shares accounted for by value advantage analysis? National market share projected? Consistent with objectives? Is all of the above articulated over the five-year plan period? Has articulated competitors' positioning and National's response?

7. *Strategic Plan*

   Did the planning team consider a realistic set of strategic alternatives? Are the Strategic Action Statements of sufficient scope and depth/detail to account for market share projections and for achieving the organization's objectives? Are they sufficiently aggressive and imaginative, especially with respect to the use of technology. Does the plan consider acquisitions or partnerships? Is there sufficient exploitation of the Company's core competencies? Does the PMP show a complete set of products? Are the lead times, ASPs and value advantages sufficiently aggressive without being unrealistic? Is there a single timeline with strategic milestones and PMP content for future review of this organization's implementation progress? Is all the above articulated over the five-year plan period.

8. *Key Assumptions*

   Are all the key assumptions on which this plan is based on the list and expressed in measurable-outcome, time-specific terms where possible? Are the owners and their commitment clearly identified to provide all required resources?

9. *Financial Summary*

   Are all the required schedules provided?

10. *Risk Analysis and Range of Results*

    Is the plan a good balance between risk and conservatism? Does the range of results capture the expected outcome to a high level of probability? Does the planning team have credible contingency plans for dealing with shortfalls?

11. *Stretch Opportunities*

    Has the team shown diligence and imagination in searching for stretch opportunities?

12. *Narrative Summary*

    Are the six questions answered adequately?

13. *The Bottom Line*

    Having listened to the plan, do you now believe these people know what they are doing, know what it takes to win, and are capable of pulling it off?

APPENDIX F

Personal and Professional Background of Gilbert F. Amelio

PERSONAL

Born March 1, 1943, in New York City (The Bronx), the first of two children.

Parents: Anthony and Elizabeth Amelio. Anthony worked in construction and in sporting goods. In WWII he served in Patton's 3rd Army and was awarded the Purple Heart. The couple are now retired in California.

Wife: Glenda Charlene Amelio; married 1989.

Children: Anthony, Tracy, and Andrew; step-children Brent and Tina Chappell. One grandchild, Logan Chappell.

Leisure Activities: Flying, skiing, tennis

EDUCATION

B.S., M.S. and Ph.D. in Physics from Georgia Institute of Technology (1965, '67, and '68)

PROFESSIONAL

In 1962, while still a student at Georgia Tech, Amelio co-founded Information Sciences, Inc., an Atlanta software firm.

He began his professional career at Bell Laboratories, assigned to the Device Development department.

In 1971 he moved to Fairchild Camera and Instrument Corporation, where he served as general manager of the Microprocessor division and also as vice president and general manager of the company's MOS Products Group.

He joined Rockwell as president of their Semiconductor Products Division (SPD) in 1983. When Rockwell realigned its telecommunications and semiconductor businesses in 1988, Dr. Amelio was made president of the new organization, Rockwell Communications Systems (RCS), headquartered in Richardson, Texas.

He became president and chief executive officer of National Semiconductor Corporation in May, 1991. In July, 1995, his responsibilities were expanded to include the position of Chairman of the Board.

Dr. Amelio has accumulated over twenty-five years of semiconductor industry experience in research, engineering, marketing, finance, product development, wafer fabrication, and assembly. Among his accomplishments are sixteen patents that he holds alone or jointly. In 1991 he was named recipient of the prestigious Masaru lbuka Consumer Electronics Award, in part a recognition of his achievements as coinventor of the industry's first charge-coupled device, which is still used today in most consumer video cameras. In addition to a great many papers published in technical journals, Amelio's article "Charge-Coupled Devices" appeared in the February, 1974, issue of *Scientific American*.

Dr. Amelio is a past member of the board of trustees of the Georgia Tech Research Corporation and past member of the university's National Advisory Board. He served as board chairman in 1985-86 during Georgia Tech's centennial year.

In 1991, he was the National Chairman of the U.S. Savings Bonds Leadership conference and 50-year anniversary celebration held in Washington, D.C.

He has also served as a member of the Executive Roundtable for the University of California Irvine and in 1993-94 as chairman of the board of directors of the Semiconductor Industry Association.

His current community and industry activities include membership in the Business Higher Education Forum and on the board of governors of the Electronics Industries Association. He is also a member of the board of directors for Apple Computer, Inc., and Chiron Corporation.

Dr. Amelio is an IEEE Fellow.

Index

D

E

F

G

H

I

K

L

M

N

O

P

Q

R

S

T

V

W

Z